BIOMETRICS, CRIME AND SECURITY

This book addresses the use of biometrics – including fingerprint identification, DNA identification and facial recognition – in the criminal justice system: balancing the need to ensure society is protected from harms, such as crime and terrorism, while also preserving individual rights. It offers a comprehensive discussion of biometric identification that includes a consideration of: basic scientific principles, their historical development, the perspectives of political philosophy, critical security and surveillance studies; but especially the relevant law, policy and regulatory issues. Developments in key jurisdictions where the technology has been implemented, including the United Kingdom, United States, Europe and Australia, are examined. This includes case studies relating to the implementation of new technology, policy, legislation, court judgments, and where available, empirical evaluations of the use of biometrics in criminal justice systems. Examples from non-western areas of the world are also considered. Accessibly written, this book will be of interest to undergraduate, postgraduate and research students, academic researchers, as well as professionals in government, security, legal and private sectors.

Marcus Smith, Adjunct Professor of Law, University of Canberra; Senior Lecturer in Law, Charles Sturt University

Monique Mann, Vice Chancellor's Research Fellow in Regulation of Technology, Faculty of Law, Queensland University of Technology

Gregor Urbas, Associate Professor of Law, Faculty of Business, Government and Law, University of Canberra

LAW, SCIENCE AND SOCIETY

Law's role has often been understood as one of implementing political decisions concerning the relationship between science and society. Increasingly, however, as our understanding of the complex dynamic between law, science and society deepens, this instrumental characterisation is seen to be inadequate, but as yet we have only a limited conception of what might take its place. If progress is to be made in our legal and scientific understanding of the problems society faces, then there needs to be space for innovative and radical thinking about law and science. Law, Science and Society is intended to provide that space.

The overarching aim of the series is to support the publication of new and groundbreaking empirical or theoretical contributions that will advance understanding between the disciplines of law, and the social, pure and applied sciences. General topics relevant to the series include studies of:

- law and the international trade in science and technology;
- risk and the regulation of science and technology;
- law, science and the environment;
- the reception of scientific discourses by law and the legal process;
- law, chaos and complexity;
- law and the brain.

Titles in this series:

Gene Editing, Law, and the Environment
Life Beyond the Human
Edited by Irus Braverman

A Socio-Legal Study of Hacking
Breaking and Remaking Law and Technology
Michael Dizon

Biometrics, Crime and Security
Marcus Smith, Monique Mann and Gregor Urbas

BIOMETRICS, CRIME AND SECURITY

Marcus Smith, Monique Mann and Gregor Urbas

Routledge
Taylor & Francis Group
a GlassHouse book

First published 2018
by Routledge
2 Park Square, Milton Park, Abingdon, Oxon OX14 4RN

and by Routledge
711 Third Avenue, New York, NY 10017

Routledge is an imprint of the Taylor & Francis Group, an informa business

British Library Cataloguing in Publication Data
A catalogue record for this book is available from the British Library

Library of Congress Cataloging in Publication Data
Names: Smith, Marcus, (Law teacher), author. | Mann, Monique, author. | Urbas, Gregor, author.
Title: Biometrics, crime and security / Marcus Smith, Monique Mann, Gregor Urbas.
Description: New York : Routledge, 2018. | Series: Law, science and society | Includes bibliographical references and index.
Identifiers: LCCN 2017038452 | ISBN 9781138742802 (hbk) | ISBN 9780815378068 (pbk)
Subjects: LCSH: Biometric identification–Law and legislation. | Evidence, Criminal. | Criminals–Identification. | Forensic sciences.
Classification: LCC K5479 .S65 2018 | DDC 363.25/8–dc23
LC record available at https://lccn.loc.gov/2017038452

A Glasshouse book

ISBN: 978-1-138-74280-2 (hbk)
ISBN: 978-0-8153-7806-8 (pbk)
ISBN: 978-1-315-18205-6 (ebk)

Typeset in Bembo
by Taylor & Francis Books

CONTENTS

ILLUSTRATIONS

Figures

Tables

1

FOUNDATIONS OF BIOMETRIC IDENTIFICATION

Introduction

The continuing advancement of scientific technology is the key factor in the vast improvements to living standards in western countries in the twentieth and twenty-first centuries. The contribution of scientific technology to crime prevention, investigation and other aspects of the criminal justice system, such as trials, is significant.

One of the earliest means of identifying individuals in the criminal justice system was fingerprinting, a technique first developed in the late eighteenth century to identify individuals based on the unique patterns on the fingertips. This technique is still used in the twenty-first century; however, it has now been digitised. Other examples of technology that has been developed to assist with criminal investigation include closed circuit television (CCTV) and deoxyribonucleic acid (DNA) identification, among others. The applications of technology in policing continue to rapidly expand into a central aspect of police work. This includes new developments relating to police information systems, big data analytics and predictive policing. All of these techniques have made very significant contributions to the investigation and prosecution of crime, but must be considered in the context of potential inaccuracy, due process implications and impacts on individual rights. Since the turn of the century, advancements in physics and information technology have provided the basis for a range of powerful identification techniques that continue to develop.

This text provides coverage and analysis of the major forms of biometrics that are currently used throughout the criminal justice sector, and, to some extent, the national security and commercial sectors. Biometrics that are presently being developed or used on a small scale, but are likely to become widely adopted in the future are also considered. The major forms of biometrics used in the criminal justice sector, including fingerprints, DNA and facial recognition will be discussed; along with emerging biometric technologies including ocular biometrics such as retina and iris recognition, voice recognition, vascular pattern recognition, keyboard dynamics, cognitive biometrics and gait analysis.

This introductory chapter is divided into four parts. The first discusses the conceptual foundations of biometric identification, and its historical development; the second examines the scientific background to some of the most commonly used biometrics; the third investigates the development of police information systems and their role in storing, searching and analysing biometric information to produce a match; and the final part considers theoretical and human rights issues that are relevant to the field of biometric identification.

Definitions

The term *biometric* refers to the measurement of a physical feature of the human body. There are many physical features that can be used as biometric identifiers. These include, most commonly, human physiology, such as patterns of the skin, aspects of the eyes, shape of the hands or blood vessel networks; facial appearance, taking account of the distance between the eyes, nose or mouth, or the general shape of the face; behavioural traits, such as gait or voice characteristics; and biodynamics, such as the pressure, pattern and speed of keystroke typing (Clarke, 1999).

As will be discussed further in later chapters, a distinction is drawn between *first generation* and *second generation* biometrics. First generation biometrics relate to physiological traits, such as fingerprint and facial recognition; while second generation biometrics include gait, keystroke analysis and cognitive biometrics.

In the context of biometric identification, a biometric feature must not only be a physiological feature capable of being measured, it must be sufficiently distinctive to form the basis of a unique identifier, capable of being efficiently verified. Biometrics should be a:

> measurable, robust and distinctive physical characteristic or personal trait that can be used to identify an individual or verify the claimed identity of an individual.
>
> *(Woodward et al., 2003, p. 1)*

Elaborating on this definition, biometrics must be capable of measurement, meaning that it must be possible to convert them to a digital format enabling database storage and searching. This has significant implications in the context of policing practice. A biometric must not change significantly over time, and be stable and reliable through the aging process, injury or changes in environmental conditions (Woodward et al., 2003).

Biometric identification can be contrasted with other means of identification, such as keys, photo identification cards and passwords that are used, for example, as proof of the right to obtain access to a restricted account, Internet site or building.

The main distinction between biometric information and other contemporary forms of identification is that it is a part of the individual themselves, rather than an object carried on the person, or a password or code. A succinct description of biometric identification used by Hopkins (1999) to distinguish it from other forms of identification is that rather than being something that an individual *knows* or *has*, it is something that they *are*.

Biometric identification involves the automatic identification of a person based on their physiological or behavioural characteristics. This is undertaken by using non-invasive computer technology to match characteristics of live individuals in real time against stored records, such as their face or fingerprint, in applications such as border or physical access control (International Biometrics Industry Association, 2000).

Historical background

The earliest known examples of biometric identification occurred in Ancient Egypt at the time of Pharaoh Khaefre (2558BC–2532BC). Biometric identification was used to identify construction workers, with the purpose of ensuring that food provided by the state was shared equitably among those legitimately eligible to receive it. Administrators developed a system that recorded the distinctive physical and behavioural characteristics of workers, along with their name, age and place of residence, to address the issues of some employees attempting to obtain more than their allocated food allowance. In the more recent past, Frenchman Alphonse Bertillon developed more scientific forms of biometric identification in the nineteenth century. Bertillon developed a method of 'judicial anthropology' which identified criminals on the basis of anatomical features (Ashbourn, 2000).

The most significant historical development in biometric identification occurred in the mid-nineteenth century. Czech scientist, Jan Evangelista Purkinje (1787–1869), established that fingerprints were unique, which had implications for identifying individuals involved in criminal activity. Research continued into the applications of fingerprinting and techniques to record them throughout the nineteenth century. Francis Galton (1822–1911), a British geneticist, first used the term *biometry* in 1901 to describe 'the application to biology of the modern methods of statistics' and made further contributions to fingerprint and facial biometric analysis (Pato & Millett, 2010, p. 17). Scotland Yard was the first law enforcement agency to use fingerprints in criminal investigations in the twentieth century, applying Galton's classification system (Ashbourn, 2000). Since that time, fingerprint identification has become an important component of criminal investigations, and remained the most important form of identifying people until the 1990s when DNA profiling became widely adopted in criminal investigations, providing a new avenue for the review of criminal convictions (Boukhonine, Krotov & Rupert, 2005).

Historically, names, passwords, personal identification numbers, keys and passwords have been used to verify individuals. However, these are proxies for the verification of a person's presence, a password can be shared, a key can be lost and a system cannot know who has presented an item (Pato & Millett, 2010). The other contemporary methods of biometric identification that are discussed throughout the text have been developed and implemented in the last two decades. They seek to apply scientific advancement to address these problems, but in turn have other issues that must be addressed.

Identification methods and issues

There are three main ways in which biometrics are used. The first is one-to-many searching, which involves a biometric profile being compared with a database of profiles to identify the individual through a direct or partial match, resulting in a computer generated likelihood that any two profiles are from the same individual. This approach is generally used in a surveillance context by police or intelligence agencies or in a criminal investigation when a DNA profile is obtained from biological material located at a crime scene and the police seek to identify an unknown individual. The second approach is one-to-one verification of identity, which seeks to determine whether an individual is who they purport to be. In this case, a live profile is provided by the individual and compared with a template stored in the computer system or identification document, such as a passport or licence. This is increasingly being used in providing an individual with physical access to a building or computer network, or to transit through international borders. The third approach in which biometrics can be used is to identify individuals on a watch-list, for example, by screening CCTV footage using facial recognition technology – a more challenging task than the methods described in the first two examples (Table 1.1) (Du, 2013).

Biometric systems involve an enrolment stage and a matching process. Enrolment involves the acquisition of the individual's biometric data, generating a template that can be stored in a database. Matching also involves the acquisition of an individual's biometric data, but adds the comparison with all templates held in a database to establish whether a match can be established (Figure 1.1) (Du, 2013).

There are several considerations in selecting physical traits for biometric identification. Each biometric has strengths and weaknesses and may be suitable for identification purposes depending on the context. There are also general suitability criteria that can be applied to each biometric identification method to assess which is the most appropriate in particular circumstances. Seven criteria have been accepted as key indicators of the suitability of biometric features. These are:

TABLE 1.1 Basic functions of biometrics

Type	*Matching*	*Question*	*Difficulty*
Verification	One-to-one	Are you who you claim to be?	Hard
Identification	One-to-many	Who are you?	Harder
Watchlist	One-to-a-few	Are you a person of interest?	Hardest

Source: Du, 2013

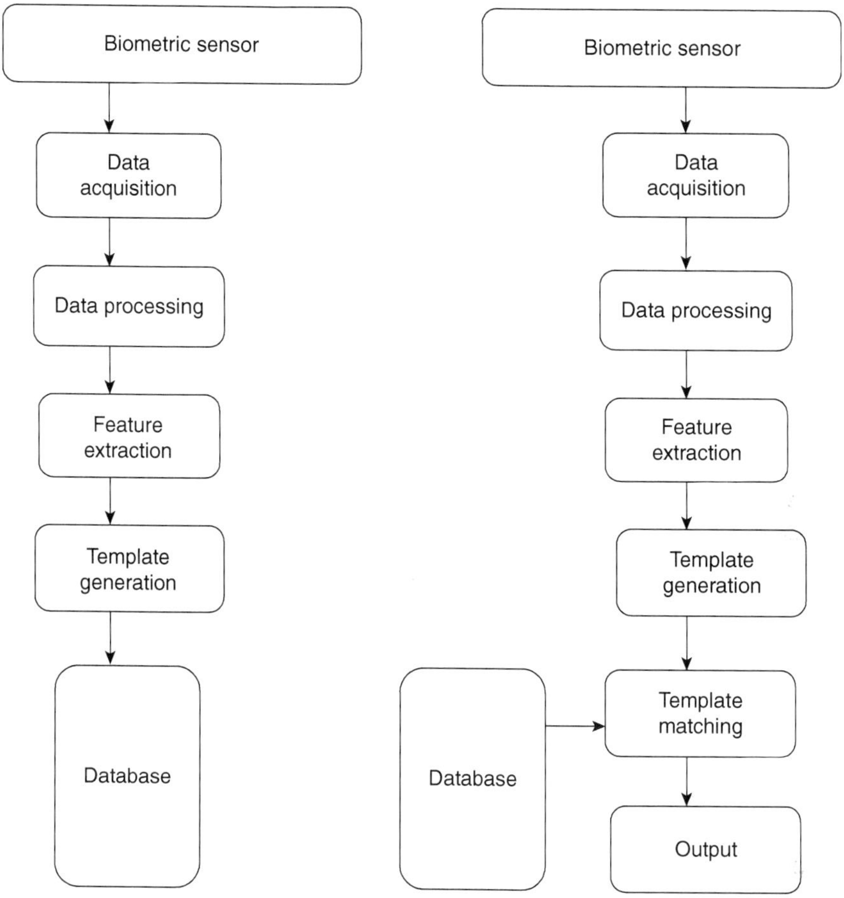

FIGURE 1.1 Biometric system enrolment and matching processes
Source: Du, 2013

universality, distinctiveness, permanence, collectability, acceptability, performance and resistance to circumvention or 'spoofing'. Collectively, this group of criteria have been referred to as the seven pillars of biometrics (Table 1.2) (Jain, Ross & Pankanti, 2006).

When compared against these criteria, certain types of biometrics have features that make them most suitable in particularly contexts. Fingerprinting may be favoured over gait analysis for accuracy in a broader range of contexts because it may be considered more distinctive and permanent, however, in some contexts (such as analysing television footage) gait analysis may be preferred because it can be assessed from a greater distance. On the other hand, in cases where a large number of people must be processed quickly, such as a border crossing, fingerprint or iris recognition may be most appropriate because a high degree of accuracy is required and a person's identity is being confirmed in close proximity.

TABLE 1.2 The seven pillars of biometrics

Universality	*Distinctive-ness*	*Permanence*	*Collectability*	*Acceptability*	*Performance*	*Resistance to circumvention*
The biometric should be present in all individuals.	The biometric feature should be sufficiently different to distinguish between individuals.	The biometric feature should be unchanged over the individual's life.	The degree of ease of collecting and measuring the biometric.	The extent to which an individual or society accepts the use of the biometric feature as a means of identification.	The degree of accuracy and the speed of the system.	The extent to which the system can be bypassed or defeated.

Source: Jain, et al., 2006

Fingerprint identification

Fingerprints are universal in the human population and remain unchanged throughout life. In combination with easy accessibility, the fact that they are deposited on surfaces by touch makes them an attractive feature for biometric identification. Fingerprints are formed in the first seven months of foetal development and are caused by the formation of nerves beneath the skin. Their key purpose is to enhance grip when handling objects. Individual fingerprints are unique among all other fingers of the same person, and indeed, among all persons, including identical twins. One limitation of fingerprint identification is that the fingerprints of approximately four per cent of the population cannot be effectively used for biometric identification purposes, due, for example, to burns and other injuries, limiting the prospect of universal population coverage. Fingerprints can also be faked relatively easily (Dessimoz et al., 2006).

Fingerprints are composed of a series of ridges and valleys in the skin on the surface of the fingertip that form a unique pattern. Fingerprint patterns are described by three key features, *arches, loops* and *whorls*: one of each is present in every fingerprint. The centre of a pattern is described as the *core* and points of discontinuity in the fingerprint ridges are known as *minutiae* (Jain, 2004). Fingerprint identification compares the unique combination of the patterns of ridges and valleys. Fingertips are placed against an optical scanner and a laser illuminates the fingerprint and converts the image into a digital format. An algorithm filters out distortions and enhances the definition of the ridges in the image (Dessimoz et al., 2006). Fingerprint identification will be discussed in detail in the next chapter.

DNA identification

DNA was first used for identification purposes in criminal investigations in the mid-1980s, and since then has made a major contribution to law enforcement around the world. Although not yet facilitating the instantaneous digital identification provided by other biometrics, DNA identification uses biotechnology to analyse a physical component of the human body (the genome) and should be considered as part of any discussion of biometric identification. The time required to undertake DNA analysis will reduce with scientific advancement and it can offer a high level of accuracy, notwithstanding the potential for human error associated with all biometrics (Smith & Mann, 2015).

DNA profiles are created by analysing the number of *short tandem repeats* that occur in specific regions of the human genome, and comprise a series of numbers. A complete match between two DNA profiles supports an inference that the samples are from the same person. However, alternatives that could also account for a match include, for example, sample contamination in the laboratory or at the crime scene. For this reason it is important that DNA evidence be considered in the context of all the available evidence in a case. DNA identification is routinely used in criminal investigations, such as a sexual assault case where the offender deposits DNA on the victim's clothing, or a homicide where the victim's hair is found on the clothing of a suspect (Smith, 2015).

Since the early 2000s, DNA databases and a range of new DNA techniques have been developed. These techniques include familial searching and mitochondrial DNA identification that can identify genetic relationships, and DNA phenotyping that can establish physical traits of an unknown suspect, such as their eye colour and ethnicity (Smith & Urbas, 2012). DNA identification is discussed in further detail in Chapter 3.

Facial recognition

Biometric facial recognition has developed relatively recently and is rapidly becoming a commonly used means of biometric identification. Facial recognition involves the creation of a template using the spatial and geometric distribution of facial features. Facial recognition compares two images using a similar algorithm to that used for digital fingerprint recognition. At an initial enrolment stage, a digital photograph of a subject's face is taken and an algorithm converts the photograph into a digital template by comparing the distances between features of the subject's face, such as the relationship of the eyes, nose, lips and chin. Where a subject's identity is sought in real time, a computer carries out the same process that took place during the enrolment stage, and compares it with an image stored in the system, to determine whether the two faces are sufficiently similar to constitute a match (Adler & Schuckers, 2007).

The image templates created by algorithms can be stored on an electronic chip within a document or identification card. Facial recognition technology has been integrated into a range of documents such as passports and identity cards, and the number of applications is likely to grow further as governments and the private sector seek to enhance the security of existing systems. Another application of facial recognition technology is its capacity to be integrated with existing television recording systems. Facial recognition technology is also less obtrusive and overt than other forms of biometrics and can be conducted at considerable distances and with low-resolution images (with varying degrees of accuracy). Facial recognition can be applied to security recording systems to screen a large population and identify persons of interest (Bowyer, Flynn & Chen, 2006). In this context, it is used by police, counterterrorism agencies, casinos and authorities at sporting and other public events to identify suspects or individuals prohibited from attending.

Facial recognition has some weaknesses in comparison with other forms of biometric identification. Facial features are subject to change as individuals age, gain or lose weight or have cosmetic procedures. Factors such as facial coverings, hairstyle, lighting, distance, rotation or movement also affect the accuracy of the results. Facial recognition is discussed in further detail in Chapter 4.

Emerging techniques and issues

New and emerging forms of biometric identification continue to be developed. Existing technologies are continually improving and becoming less expensive, which facilitates wider adoption. Chapter 5 foreshadows the introduction of the most developed and widely implemented new and emerging biometric modalities, and discusses issues associated with accuracy, such as *spoofing*. These include new developments in physiological forms of identification, including ear recognition, vascular pattern recognition, ocular biometrics, voice recognition, gait recognition, keystroke dynamics and cognitive biometrics. In turn, each of these has a range of possible applications in crime and security and associated advantages and disadvantages.

Although biometric identification techniques are considered to be less vulnerable to fraud or forgery than other forms of identification, they are not without vulnerabilities. Spoofing attacks are a major problem for the successful deployment of biometric systems. Attacks on biometric systems and data breaches are significant as biometric traits often correspond to an individual's physical existence and when compromised they cannot be deleted or replaced. There is the risk of the leakage or hacking of stored biometric information from databases or electronic chips. This is significant to consider as biometric identification relies on the accuracy and reliability of identification (Chingovska, dos Anjos & Marcel, 2014). Spoofing refers to an attempt to gain unauthorised access or defeat a biometric system through either direct or indirect attacks. There are two main types of spoofing attacks: direct attacks that target the input of the sensor of the biometric system and indirect attacks that target the inner workings of the system (Rebera, Bonfanti & Venier, 2014). Biometric systems can be compromised by the exploitation of security gaps (Toli & Preneel, 2015) or the use of artificial replicates to falsify identity. Examples of the latter include the use of gelatine clones of fingerprints, contact lenses that include an individual's iris pattern or mimicking behaviour biometrics (Toli & Preneel, 2015). All physical biometric modalities are vulnerable to spoofing as they can be replicated (with varying degrees of difficulty), and behavioural biometrics can be mimicked (Rebera et al., 2014).

Most commentary focuses on spoofing at the direct sensor level via the use of artificial biometric samples or manipulation of identity (Rebera et al., 2014). This involves submitting a false replica of the biometric sample to the sensor. For direct attacks at the sensor level, it is impossible to use encryption or digital signatures because the attack is directed outside the digital limits of the biometric system. This can also involve the manipulation of stored biometric templates or exploitation of error rates (Biggio et al., 2011).

There is a long history of attempts to spoof fingerprints. An early case involved a prisoner in a Kansas prison in the 1920s named Alert Wehde who used his expertise in photography and engraving to produce fake fingerprints. Wehde took photographs of latent fingerprints and then used the photographs to etch the fingerprints on copper plates that were then used to create fake fingerprints (Biggio et al., 2011). It is possible to create fingerprint moulds from an individual's finger or fingerprints left on a surface (Schuckers, 2002). More recent attempts involve defeating fingerprint security in smart phones using similar techniques. For example:

> Two days after the Apple iPhone 5s appeared with its new TouchID fingerprint security in September 2013, the Hamburg based Chaos Computer Club (CCC) managed to hack the phone's fingerprint reader using a latex model of a print taken from the 'victim' created on a 3D printer. Later that year the CCC even managed to copy fingerprints from a photo of a victim's hand, avoiding the need to lift their actual prints.
>
> *(Ring, 2015, p. 5)*

Empirical research has shown that fake fingerprints are highly effective at defeating biometric systems. For example, Matsumoto et al. (2002) tested a range of commercial fingerprint sensors with fake fingerprints with a success rate of higher than 60 per cent. Other research has produced success rates of over 70 per cent by using different methods and materials to spoof fingerprints (Biggio et al., 2011).

Police information systems

Biometric templates are an increasingly significant part of the suite of police information systems around the world. The digitisation of information that has occurred since the 1980s has contributed to significant increases in the volumes of data stored and the efficacy of data searching, matching and management. As with many other areas of government and business, it has changed the way police agencies approach criminal investigation (Byrne & Marx, 2011). Historically, police information systems comprised paper-based file and index catalogue systems. These required a large amount of storage space and were time-consuming to interrogate in order to identify a possible match. They also allowed little scope for information sharing outside specific jurisdictions or commands, and in the event information was shared, this was likely to be onerous and time-consuming. Advancements in information technology over the past 30 years have enabled police agencies to increase the exchange and dissemination of information. Developed countries have moved towards national databases, and in recent years, to transnational information sharing (Luen & Suliman, 2001).

Databases are used in police contexts to store and compare information about crime scenes, individuals and networks (Varano et al., 2007). They range from basic record management systems, to complex analytical software systems that have the potential to inform tactical and strategic intelligence, and databases of human biometrics (Ratcliffe, 2016). Potential benefits of the use of these databases include increased efficiency of information sharing (Koper, Lum & Willis, 2014), and in particular, cross-jurisdictional criminal investigation (Nuth, 2008). While publicly available data on the impact these databases have on investigation outcomes is limited due to sensitivities associated with the nature of the information, some research has found that the introduction of new technology to police agencies does not always produce anticipated improvements in productivity, communication or management, and that the introduction of new technology creates additional administrative work without actually contributing to greater crime reduction (Koper et al., 2014).

Police information systems have the potential to improve policing through the analysis of data without the need for human intervention. Databases can improve the speed of crime detection, and assist in the strategic planning of policing (Koper et al., 2014). Law enforcement databases may enhance cross-jurisdictional cooperation and coordination, as individual police organisations maintain their own information base, while receiving shared information on relevant matters (Dunworth, 2000).

National databases

In the United States, the Science and Technology Branch of the Federal Bureau of Investigation (FBI) is responsible for the development and maintenance of national police information systems. The Criminal Justice Information System (CJIS), created in February 1992, is the central repository of criminal justice information, for the FBI and the other US federal, state and local law enforcement agencies. US databases include the National Crime Information Center (NCIC), the National Instant Criminal Background Check System (NICBCS), the Combined DNA Index System (CODIS) and the National Integrated Ballistics Information Network (NIBIN) (Federal Bureau of Investigation, 2017b).

In the United Kingdom, the Home Office and the Association for Police and Crime Commissioners manage Britain's police information systems (Association of Police and Crime Commissioners, 2015). Current databases include the Police National Database (PND), the Police National Computer (PNC), the National DNA Database (NDNAD), the National fingerprint and identity platform database (IDENT1) and the National Ballistics Intelligence Services (NABIS) (National Police Improvement Agency, 2013). Significantly, the United Kingdom has created a Commissioner for the Retention and Use of Biometric Material to ensure there was an office responsible for governing the retention and use of biometric information (currently only DNA or fingerprints fall within the Commissioner's authority). The Biometrics Commissioner regulates the use of biometric information, provides protection from disproportionate enforcement action and limits the application of surveillance and counter-terrorism powers (*Protection of Freedoms Act 2012* (UK)).

The Australian Criminal Intelligence Commission (ACIC) was formed in 2016 following a merger between the Australian Crime Commission (ACC) and the CrimTrac Agency. CrimTrac had been responsible for the development, sharing and maintenance of law enforcement databases in Australia since July 2000, while the ACC was a federal agency established to investigate organised crime. According to the ACIC, its databases seek to enhance Australian policing and law enforcement, and 'contribute directly to the effectiveness and efficiency of police and law enforcement agencies in Australia' (ACIC, 2017). In addition to DNA and fingerprints, the ACIC administers national databases relating to ballistics, cybercrime reports, firearms ownership, vehicles and persons of interest (ACIC, 2017).

It can be expected that government biometric databases, initially established to identify citizens for social security payments and other government services, will also be available to police for criminal investigations. In India, the Unique Identification Authority operates the largest biometric database in the world, currently including 700 million citizens' iris and fingerprint templates as well as demographic information. This information is included on a national identity card, known as an Aadhaar card. It is increasingly being used by police in criminal investigations, and is now required when a complainant reports a crime to police (Mitra & Gofman, 2016).

Database impact

Measuring the impact of modern law enforcement databases is challenging and the empirical evidence is limited (Nuth, 2008). The research that is available contains mixed findings regarding effectiveness, case resolution and overall reductions in crime (Koper et al., 2014). This section examines the empirical evidence in relation to the impact of information systems on policing and police outcomes that are relevant to biometric identification.

The most widely discussed aspect of law enforcement databases is their potential to improve the efficiency and effectiveness of investigations. Through a series of interviews and focus groups Chan investigated the extent to which information databases have modified the practice of policing. The study found that police databases enable police officers to work more effectively, cope with a large volume of policing related information, share information effectively and work more cooperatively (Chan, 2001). Other studies have found the impact of law enforcement databases is limited, but that they offer significant improvement to police performance, communication and information sharing (Byrne and Marx 2011), as well as improving the investigative process by assisting information flow (Koper, Lum & Willis, 2015). Conversely, other research has found that databases offer little or no significant improvements in police performance. More broadly, Harris (2007) found that police information systems (including record management systems, information-sharing systems, computer-aided dispatch systems and crime mapping systems) had no significant impact on policing. Hekim, Gul & Akcam (2013) examined the use of police databases for criminal investigative purposes in 233 law enforcement agencies in the United States and found no significant relationship between case clearance rates (calculated by dividing the number of crimes cleared – with a charge being laid – by the total number of crimes that are recorded) and the use of police databases.

A range of issues can impact the effectiveness of police information systems such as poor implementation and underutilisation of the databases, as well as a lack of training (Koper et al., 2015). Data security, missing or inaccurate data (completeness and validity), siloed information, ineffective human-computer interfaces, poor search capabilities and hardware limits need to be considered and managed when implementing new information systems into police agencies and practices (Koper et al., 2014). Research conducted in the United States found that ineffective user interfaces, loss of connectivity, loss of data and technological literacy had a negative effect on both police attitudes and performance (Koper et al., 2015). The culture of law enforcement agencies is often politically and organisationally conservative, and this has been described by some as an impediment to change (Chan, 1996). The introduction of new technologies that influence police practices can introduce new accountability requirements, reduce communication, and produce resistance. Koper et al. (2015) found that aspects of police organisational culture include resistance to change and collaboration, and an emphasis on traditional reactive policing tasks. This research found that police officers believed technology detracted from the 'important' aspects of policing, including interacting with people, and developing good situational awareness (Koper et al., 2015). Cultural resistance, resentment of accountability demands and limits to discretion may lead police officers to manipulate databases to align with more traditional aspects of policing (Harris, 2007). Over time, with new generations of police, improved technology and the increased establishment of biometric databases, this may improve. However, given the investment being made in biometric police information systems, and their expansion, further research to establish their impact on investigations would be beneficial.

Theoretical perspectives

Theoretical perspectives provide a framework for understanding the rationale for the introduction and use of biometric technology and a basis for understanding whether this is appropriate in certain circumstances. This may form part of a consideration as to whether expenditure in implementing these systems is justified, whether it is appropriate or permissible to require citizens to submit their biometrics, and for those to be used in criminal investigations. Investigating and applying relevant theories is an important step in developing a considered position on the adoption of biometric identification technology, including when and how it should be implemented. This discussion describes key criminological and political theories that assist in decision-making that seeks to balance the security of society and the rights of individuals.

Rational choice and situational crime prevention

The use of biometric information to prevent crimes (for example, by restricting access), is a key application of biometrics. Situational crime prevention is an extension of rational choice theory to understand how opportunity structures can be manipulated to prevent individuals from committing crime. Crime is understood to result from choices made by offenders on the basis of a calculation of the risks and rewards of these choices (Clarke, 1997). Rather than focusing on the psychological or sociological background of an individual, as is the case with many criminological theories, this approach is concerned with rational decision-making in the context of the immediate situational dynamics.

In order to reduce the number of opportunities to commit crime, several factors must be taken into account. These include: the characteristics of the places and situations that are potentially exposed to criminal activities; the aspects that draw potential criminals towards these places and situations; the way in which potential criminals are able to take advantage of the opportunities at these places and situations; and the immediate factors that lead to criminal behaviour. Situational crime prevention asserts that crime can be prevented if potential targets are securely guarded, the ability to commit crime is controlled and potential criminals are monitored (Siegel, 2011).

Historically, there have been three main criminological theorists associated with crime prevention. Oscar Newman developed the concept of *defensible space*, arguing that crime can be prevented or reduced through architectural designs that reduce opportunity, such as lighting that enhances surveillance (Newman, 1972). C. Ray Jeffery extended Newman's ideas, describing how mechanisms, such as security systems and neighbourhood watch programmes, can reduce opportunities to commit crime (Jeffery, 1977). More recently, Ronald Clarke has compiled strategies of crime prevention that can be used in combination to create an environment that is not conducive to crime (Clarke, 1997).

Biometric techniques can form a basis for crime prevention that is similar to more traditional forms of target hardening, such as photographic identity cards, passports, locks and keys (access control) and pin numbers, to increase the effort required to commit crimes. Biometric identification applies new technology and modernises established forms of crime prevention. Cornish and Clarke (2003) describe the key aspects of situational crime prevention: increasing the efforts and risks of committing crimes and reducing the rewards, provocations and excuses for doing so. Increasing the effort needed to commit crime is relevant in the context of biometrics, as well as many others, including the capacity of biometrics to reduce anonymity, strengthen formal surveillance, discourage imitation and assist compliance. The potential relevance of biometric identification to techniques of situational crime prevention is outlined in a table adapted from one proposed by crime prevention theorists Table 1.3.

TABLE 1.3 Applying Situational Crime Prevention to Biometric Identification (BI)

Increase the effort	*Increase the risks*	*Reduce the rewards*	*Reduce provocations*	*Remove the excuses*
1. Harden targets BI may provide better security than traditional methods such as photos and pin numbers.	***6. Extend guardianship*** BI extends guardianship via surveillance.	***11. Conceal targets***	***16. Reduce frustration and stress***	***21. Set rules*** Require individuals to provide BI.
2. Control access to facilities BI can more securely control access than traditional methods such as keys and swipe cards.	***7. Assist natural surveillance*** BI CCTV integration can provide surveillance of public places	***12. Remove targets***	***17. Avoid disputes***	***22. Post instructions***
3. Screen exits BI is suitable for both entry and exit screening purposes.	***8. Reduce anonymity*** BI reduces anonymity.	***13. Identify property*** BI can more certainly establish ownership.	***18. Reduce emotional arousal***	***23. Alert conscience*** Require individuals to provide BI
4. Deflect offenders BI provides greater security and can increase deterrence. Increased risk of detection may also influence decision-making.	***9. Utilise place managers*** BI systems increase availability of human resources to oversight.	***14. Disrupt markets*** BI makes human trafficking more difficult by making it harder to mask the identity of victims.	***19. Neutralise peer pressure***	***24. Assist compliance***
5. Control tools/ weapons BI can enhance control of sensitive information and equipment.	***10. Strengthen formal surveillance*** BI formal surveillance.	***15. Deny benefits***	***20. Discourage imitation***	***25. Control drugs/ alcohol***

Source: Adapted from Cornish and Clarke, 2003

Freedom and security

Social contract theory is a central political theory used to justify the creation of a state and the obligations of citizens, providing the rules necessary to ensure that society can function effectively. An appreciation of contract theory begins with an understanding of the concepts of the *state of nature* and the *social contract*. The state of nature describes the condition of humans prior to government being established. It highlights the advantages of political organisation, and of implementing some form of governmental authority to improve living conditions. The social contract hypothetically describes how individuals without government form a society and accept obligations to each other, and to the state. The social contract forms the basis of law in the state, and justifies individual obligations.

According to Thomas Hobbes (1588–1679), humans are naturally egoistic, and as they cannot otherwise pursue their goals without conflicting with one another, individuals must relinquish their natural rights and become subjects of the state. The state acquires *de jure* authority and becomes the 'great Leviathan' allowing individuals to construct a stable society, with the state obliged to protect its citizens (Hobbes, 1660). John Locke develops this line of argument further, arguing that the power of the government is not absolute, but limited to upholding the liberty, life and property of the individual (Locke, 1690).

There are a range of circumstances where citizens are required to concede some of their rights to the government, often because the state has far greater resources and can provide important services that improve living standards. It can be argued that the government retaining biometric information about private citizens and requiring that citizens submit to biometric identification procedures occurs as part of a social contract between the government and citizens. Although some may argue that the use of individuals' bodies in this manner is an unwarranted infringement of individual rights, others hold a view that the inclusion of citizens' biometric information in a national database would be necessary to ensure that the police can prevent crime and identify individuals responsible for committing crimes, that the government can effectively protect its borders and that citizens can enjoy security and freedom.

Cases where the state invades the liberties of citizens include: collecting personal information from individuals in order to use government services; retaining photographs and biographical information on registered drivers; retaining the medical histories of citizens using public healthcare services; and retaining details of income and expenses through the taxation system while deducting a significant proportion to build and support public infrastructure and services. These examples are invasive and infringe individual liberties to some extent, however, they are tolerated because the provision of benefits is thought to meet or exceed the costs to individual citizens. Similarly, the requirement that the government retain unique details about an individual's retina, fingerprints or facial features for the purpose of upholding public safety could be considered no more invasive than many of the existing measures described above, and would be a reasonable concession in light of the benefits that can be obtained by ensuring that the government can make the best use of available technology in carrying out its key function to ensure security.

The second political theory that will be considered is *utilitarianism*. According to it, all human actions, inactions and policies have consequences for the individual and others. A utilitarian would argue that society is best arranged so that it achieves the greatest overall satisfaction when all of its members are considered. Jeremy Bentham (1748–1832) and John Stuart Mill (1806–1873) developed utilitarian theory in the seventeenth century. Bentham described utility as 'that property in any object, whereby it tends to produce benefit, advantage, pleasure, good or happiness' (Bentham, 1798, p. 2). In Bentham's view, the best way to govern a society is to provide the most pleasure and the least pain for citizens:

> that principle which approves or disapproves of every action whatsoever, according to the tendency which it appears to have to augment or diminish the happiness of the party whose interest is in question … I say of every action whatsoever; and therefore not only of every action of a private individual, but of every measure of government.
>
> *(Bentham, 1798, p. 2)*

Utilitarianism provides a framework that can be used to evaluate whether potential costs, such as privacy, are outweighed by the benefits, such as potential improvements to public safety and security.

There have been a number of influential accounts of human freedom, or liberty. *Liberty*, as defined by John Stuart Mill, is 'the nature and limits of the power which can be legitimately exercised by society over the individual' (Mill, 1859, p. 1). Although political freedom is associated with the recognition of individual rights and constitutional safeguards, some encroachment upon individual rights is in the public interest. For instance, Mill states that:

> The sole end for which mankind are warranted, individually or collectively, in interfering with the liberty of action of any of their number, is self-protection. That the only purpose for which power can be rightfully exercised over any member of a civilized community, against his will, is to prevent harm to others … Over himself, over his own body and mind, the individual is sovereign.
>
> *(Mill, 1859, p. 1)*

Further, though Mill emphasises the importance of non-interference, he states that individual autonomy may be limited if it is likely that harmful consequences to other individuals may result: 'That the only purpose for which power can be rightfully exercised over any member of a civilised community, against his will, is to prevent harm to others' (Mill, 1859, p. 1).

Libertarians may object to the wide-scale adoption of biometric technology, on the basis that utilising aspects of a person's body for identification in this manner violates an individual's right to privacy. However, in general terms, utilising biometric identification technology fits within Mill's guidelines. It can potentially contribute to better prevention and policing of crime, and therefore, the prevention of harm to others and increased security. There are numerous ways in which police and the government currently identify individuals and breach privacy in ways that are accepted due to the overall benefits they provide.

John Rawls (1921–2002) sought to establish principles to ensure liberty, equality and justice in society, describing his model as 'justice as fairness' (1971, p. 2). According to Rawls, the decision-makers should adopt a hypothetical *original position* where factors such as social status, wealth, gender and ethnicity are unknown. A decision-maker in such a position should be motivated by concern for all persons in society, irrespective of their position (Rawls, 1971, p. 24).

In the context of the adoption of biometric technology in passports, for example, the requirement to submit biometric data applies equally to all citizens; from those holding the highest political and commercial offices in the state, to those performing unskilled labour. It would not lead to the over-representation of particular racial groups, and could actually reduce perceptions of racial discrimination, for example by border control officers. The universal adoption of biometric identification measures offers inherent equality.

However, the volume of information collected by governments for security purposes increased significantly following the attacks on 11 September 2001. Biometric identification technology and other recent developments such as metadata retention can provide detailed insights into citizens' lives, and the use of human biometric information in law enforcement investigations and other applications continues to expand. Biometric technologies are being integrated with other government information sources, such as metadata, CCTV and social media. Information sharing arrangements that facilitate searches across state, national and transnational government databases are also expanding. It is commonly observed that government law and regulation lag behind technological advancements. In some countries, such as the United States, there is a constitutional bill of rights or a cause of action for serious invasion of privacy; in other countries, such as Australia, there are limited statutory protections available in response to the expansion of biometric information repositories (Mann & Smith, 2017).

The main privacy concerns associated with biometrics relate to the circumstances in which biometric information is obtained, retained, stored and shared between agencies; as well as the overall purposes for which it is used (de Andrade, Martin & Monteleone, 2013). Biometric technology can be considered invasive as it identifies individuals and can be used to link and connect information across datasets (De Hert, 2013). There are a range of privacy interests at stake with respect to biometric information. These vary according to a number of factors, for example, whether they are used for verification (one-to-one confirmation) or identification (one-to-many database search), whether identifiable data or templates are stored and shared and whether information is stored in a centralised database or localised device (Campisi, 2013). Although these types of considerations and potential privacy impacts are relevant to all forms of biometric information, they are especially important in the context of biometrics such as facial recognition technology, because faces are difficult to hide and alter, and are linked to an individual's physical existence (de Andrade et al., 2013). This biometric presents privacy considerations as it can be used to locate and track individuals through widely implemented CCTV surveillance systems, as discussed above.

The legal and philosophical concept of privacy is the assertion that some aspects of an individual's life are personal and should be free from intrusion (Warren & Brandeis, 1890). Scholars have argued that the consequence of this balancing approach is that 'individual rights are invariably traded off against the community interests in preventing, detecting and prosecuting crime' (Bronitt & Stellios, 2005, p. 887). These exemptions, particularly when coupled with no constitutional bill of rights, privacy legislation or an enforceable privacy tort in a particular jurisdiction, would demonstrate limited privacy protections against the expansion of community interests (de Zwart, Humphreys & van Dissel, 2014).

One prominent concern about the inadequacy of privacy protections is the potential for 'function creep', where information taken for a particular purpose is used for other purposes for which consent was not obtained (Brey, 2004). An example of this would be the creation of a national database of facial templates created using driver licence or passport photographs. This may be an example of function creep because individuals consented to providing a photograph to obtain a passport, but did not consent to their biometric information being extracted from that image and being used for law enforcement, security or intelligence purposes. Although photographs have been a resource available for use in police investigations for some time (Edmond & San Roque, 2014), the scale, digitisation, automation and integration of information provided by biometric facial recognition could be considered a shift in the way that the photographs are used, and may warrant a more detailed consideration.

References

Adler, A. & Schuckers, M. (2007). Comparing human and automatic face recognition performance. *IEEE Transactions on Systems, Man, and Cybernetics – Part B: Cybernetics* 1248.

Ashbourn, J. (2000). *Biometrics: Advanced Identity Verification*. London: Springer-Verlag.

Association of Police and Crime Commissioners. (2015). PCCs to establish ground-breaking national police ICT company. Retrieved from http://apccs.police.uk/press_release/pccs-establish-ground-breaking-national-police-ict-company

Australian Criminal Intelligence Commission. (2017). *Biometric Matching*. Retrieved from https://www.acic.gov.au/our-services/biometric-matching

Australian Law Reform Commission. (2008). *For Your Information: Australian Privacy Law and Practice, Report No 108*. Canberra: Australian Government.

Bentham, J. (1798). *An Introduction to the Principles of Morals and Legislation*. Whitefish, MT: Kissinger.

Biggio, B., Akhtar, Z., Fumera, G., Marcialis, G. & Roli, F. (2011). Security evaluation of biometric authentication systems under real spoofing attacks. *Institute of Engineering and Technology Biometrics 1*(1), 11.

Boukhonine, S., Krotov, V. & Rupert, B. (2005). Future security approaches and biometrics. *Communications of the Association for Information Systems 16*, 937.

Bowyer, K., Flynn, P. & Chen, X. (2006).Face recognition using 2-D, 3-D, and infrared: Is multimodal better than multisample? *Proceedings of the IEEE 94*.

Brey, P. (2004). Ethical aspects of facial recognition systems in public places. *Journal of Information, Communication and Ethics in Society 2*, 97.

Byrne, J. & Marx, G. (2011). Technological innovations in crime prevention and policing: A review of the research on implementation and impact. *Journal of Police Studies 20*, 39.

Campisi, P. (2013). Security and privacy in biometrics: Towards a holistic approach. *In* Campisi, P. (ed.), *Security and Privacy in Biometrics*. London: Springer-Verlag.

Chan, J. (1996). Changing police culture. *British Journal of Criminology 36*, 109.

Chan, J. (2001). The technological game: How information technology is transforming police practice. *Criminal Justice 1*, 139.

Chingovska, I., dos Anjos, A. & Marcel, S. (2014). Biometrics evaluation under spoofing attacks. *IEEE Transactions on Information Forensics and Security 9*(12), 2264.

Clarke, R. (1995). Situational crime prevention. *In* Tonry, M. & Farrington, D. (eds), *Building a Safer Society: Strategic Approaches To Crime Prevention*. Chicago: University of Chicago Press.

Clarke, R. (1997). *The Australian Privacy Act 1988 as an Implementation of the OECD Data Protection Guidelines*. Retrieved from http://www.rogerclarke.com/DV/PActOECD.html

Clarke, R. (1999). *Introduction to Dataveillance and Information Privacy and Definitions of Terms*. Retrieved from http://www.rogerclarke.com/DV/Intro.html

Cornish, D. & Clarke, R. (2003). Opportunities, precipitators, and criminal decisions: A reply to Wortley's critique of situational crime prevention. *In* Smith, M. & Cornish, D. (eds), *Theory for Practice in Situational Crime Prevention* (pp. 41–96). Monsey, NY: Criminal Justice Press.

de Andrade, N., Martin, A. & Monteleone, S. (2013). All the better to see you with, my dear: Facial recognition and privacy in online social networks. *IEEE Security and Privacy 11*(3), 21.

De Hert, P. (2013). Biometrics and the challenge to human rights in Europe: Need for regulation and regulatory distinctions. *In* Campisi, Patrizio (ed.), *Security and Privacy in Biometrics*. London: Springer.

Dessimoz, D., Richiardi, J., Champod, C. & Drygajlo, A. (2006). *Multimodal Biometrics for Identity Documents: State-of-the-Art*. Lausanne: University of Lausanne Press.

de Zwart, M., Humphreys, S., & van Dissel, B. (2014). Surveillance, big data and democracy: Lessons for Australia from the US and UK. *University of New South Wales Law Journal 37*, 713.

Du, E. (2013). *Biometrics: From Fiction to Practice*. Singapore: Pan Stanford.

Dunworth, T. (2000). Criminal justice and the IT revolution. *In* Horney, J., Mackenzie, D., Martin, J. Peterson, R. & Rosenbaum, D. (eds), *Policies, Processes and Decisions in the Criminal Justice System*. Washington, DC: National Institute of Justice.

Edmond, G. & San Roque, M. (2014). Honeysett v The Queen: Forensic science, "specialised knowledge" and the uniform evidence law. *Sydney Law Review 36*, 323.

Federal Bureau of Investigation (FBI). (2017). *Criminal Justice Information Services*. Retrieved from http://www.fbi.gov/about-us/cjis

Federal Bureau of Investigation (FBI). (2017). *Laboratory Services*. Retrieved from http://www.fbi.gov/about-us/lab

Harris, C. (2007). The police and soft technology: How information technology contributes to police decision making. *In* Byrne, J. & Rebovich, D. (eds), *The New Technology of Crime, Law and Social Control*. Monsey, NY: Criminal Justice Press.

Hekim, H., Gul, S. & Akcam, B. (2013). Police use of information technologies in criminal investigations. *European Scientific Journal 9*, 221.

Hobbes, T. (1660). *Leviathan*. London: Continuum.

Hopkins, R. (1999). An introduction to biometrics and large scale civilian identification. *International Review of Law, Computers and Technology 13*, 337.

International Biometrics Industry Association. (2000). *Interest in Biometric Industry Continues to Soar*. Retrieved from http://www.ibia.org

Jain, A. (2004). An introduction to biometric recognition. *IEEE Transactions on Circuits and Systems for Video Technology 14*, 1.

Jain, A., Ross, A. & Pankanti, S. (2006). Biometrics: A tool for information security. *IEEE Transactions on Information Forensics and Security 1*, 125.

Jeffery, C. (1977). *Crime Prevention through Environmental Design*. Beverly Hills, CA: Sage.

Koper, C., Lum, C., & Willis, J. (2014). Optimizing the use of technology in policing: Results and implications from a multi-site study of the social, organizational, and behavioural aspects of implementing police technologies. *Policing 8*, 212.

Koper, C., Lum, C., & Willis, J. (2015). *Realizing the Potential of Technology in Policing. Fairfax County*. Fairfax, VI: George Mason University Press.

Locke, J. (1690). *Second Treatise of Government*. Whitefish, MT: Kessinger.

Luen, T., & Suliman, A. (2001). Knowledge management in the public sector: Principles and practices in police work. *Journal of Information Science 27*, 311.

MacKenzie, D., Martin, J., & Rosenbaum, D. (eds) (2000). *Policies, Processes and Decisions of the Criminal Justice System*. Washington, DC: National Institute of Justice.

Mann, M. & Smith, M. (2017). Automated facial recognition technology: Recent developments and approaches to oversight. *University of New South Wales Law Journal 40*, 121.

Matsumoto, T., Matsumoto, H., Yamada, K. & Hoshino, S. (2002). Impact of artificial 'gummy' fingers on fingerprint systems. *Proceedings of the SPIE 4677*, 275.

Mill, J. (1859). On liberty. *In* Gray, D. (ed.) (1998), *On Liberty and Other Essays*. New York: Oxford University Press.

Mitra, S. & Gofman, M. (2016). *Biometrics in a Data Driven World: Trends, Technologies, and Challenges*. Boca Raton, FL: Taylor & Francis.

National Police Improvement Agency. (2013). *National Information Services*. Retrieved from http://www.npia.police.uk/en/19483.htm

Newman, O. (1972). *Defensible Space: Crime Prevention through Urban Design*. New York: Macmillan.

Nuth, M. S. (2008). Taking advantage of new technologies: For and against crime. *Computer Law & Security Report 28*, 437.

Pato, J. & Millett, L. (2010). *Biometric Recognition: Challenges and Opportunities*. Washington, DC: National Academies Press.

Ratcliffe, J. H. (2016). *Intelligence-led Policing* (2nd edition). London: Routledge.

Rawls, J. (1971). *A Theory of Justice*. Boston: Harvard University Press.

Rebera, A., Bonfanti, M. & Venier, S. (2014). Societal and ethical implications of anti-spoofing technologies in biometrics. *Science and Engineering Ethics 20*, 155.

Ring, T. (2015). Spoofing: Are the hackers beating biometrics? *Biometric Technology Today 7*, 5.

Schuckers, S. (2002). Spoofing and anti-spoofing measures. *Information Security Technical Report 7*, 56.

Siegel, L. (2011). *Criminology* (11th edition). Belmont, CA: Wadsworth.

Smith, M. (2015). *DNA Evidence in the Australian Legal System*. Sydney: Lexis Nexis.

Smith, M. & Mann, M. (2015). Recent developments in DNA evidence. *Trends and Issues in Crime and Criminal Justice No. 506*. Canberra: Australian Institute of Criminology.

Smith, M. & Urbas, G. (2012). Regulating new forms of forensic DNA profiling under Australian legislation: Familial matching and DNA phenotyping. *Australian Journal of Forensic Sciences 44*, 63.

Toli, C. & Preneel, B. (2015). Provoking security: Spoofing attacks against crypto-biometric systems. In Shoniregun, C. & Marksheffel, B. (eds), *IEEE World Congress on Internet Security (WorldCIS-2015)*. Red Hook, NY: Curran & Associates Inc.

Varano, S., Cancino, J., Glass, J., & Enriquez, J. (2007). Police information systems. *In* Schafer, J. (ed), *Policing 2020*. Washington, DC: Federal Bureau of Investigation.

Warren, S. & Brandeis, L. (1890). The right to privacy. *Harvard Law Review 4*, 193.

Woodward, J., Horn, C., Gatune, J., & Thomas, A. (2003). Biometrics: A look at facial recognition. *RAND Documented Briefing*. Santa Monica, CA: RAND.

2

FINGERPRINT BIOMETRICS

Introduction

The subtle ridges and valleys of the skin on the tips of the fingers were the first form of biometric to be developed for systematic human identification. In the early twentieth century, fingerprints were obtained by rolling an inked finger on paper. More recently, electronic sensor technology has enabled fingerprints to be recorded digitally, along with the capacity for comparison via automated analysis. This chapter will examine the development and application of digital fingerprint identification in a range of contexts, including law enforcement, border security and, to some extent, the private sector. The chapter begins with a discussion of the scientific and historical development of fingerprint identification, tracing its advancement to its status today, as arguably the most widely used and accepted form of biometric identification in operation.

The discussion will include the establishment of fingerprint databases in a law enforcement setting, and then in broader contexts that have developed more recently, such as border security purposes. The development of fingerprint databases in key jurisdictions around the world will be outlined, providing an understanding of their scale and application. The final section of this chapter will discuss broader applications of fingerprint identification, focusing on its use in criminal investigations, border security, as well as more recent integration in the private sector, such as the banking, communications and firearms industries. Current issues, and potential future developments that will be covered, include accuracy, security and capacity to be used as evidence. The chapter will include a discussion of key developments, including the impact of fingerprint identification systems established around the world, particularly the United States, United Kingdom and Australia.

Scientific and historical development

Fingerprints are created by the formation of nerves beneath the skin and also have an important functional purpose: they assist with grip when handling objects. Fingerprints are unique between all fingers of the same person, and indeed among all people, including identical twins (Jain, Ross & Prabhakar, 2004).

Human fingerprints are composed of a series of ridges and valleys in the skin on the surface of the fingertip. These ridges and valleys form the unique pattern of a fingerprint (Jain et al., 2004). The architecture of a fingerprint can be classified into a series of *arches, loops* and *whorls* (Figure 2.1). At least one of each of these features is present in every fingerprint. Key terminology used to describe the differences between individual fingerprints includes: the *core*, the term used to describe the centre of a pattern; and the *delta*, the point from which three patterns deviate. *Minutiae* is the name used for points of discontinuity in the fingerprint ridges, and these can be divided into *endings* and *bifurcations* (Figure 2.2) (O'Gorman, 1999).

Fingerprints can be categorised into three types: visible fingerprints, latent fingerprints and plastic fingerprints. Visible fingerprints are clearly discernible to the naked eye, and are formed when a substance, such as blood or ink, is transferred onto a surface when pressure is applied by a finger. Latent fingerprints are not visible to the naked eye, and require further enhancement by forensic scientists before they can be visualised. Traditionally, this enhancement is in the form of a chemical or powder visualisation technique that highlights the architecture of the fingerprint, in order to allow a photograph or impression to be taken for analysis and comparison. The third category of fingerprints, plastic fingerprints, are formed when a negative impression is created from pressure applied by a finger to a delicate material. Examples of materials likely to form a plastic fingerprint include silicone, wax and clay (Saferstein, 2015). This type of fingerprint could be used to create a mould and a replica fingerprint that may be used in an attempt to imitate another person to gain access to a system, as discussed in Chapter 1.

FIGURE 2.1 Fingerprint features
Source: Copyright 2008 L. O'Gorman

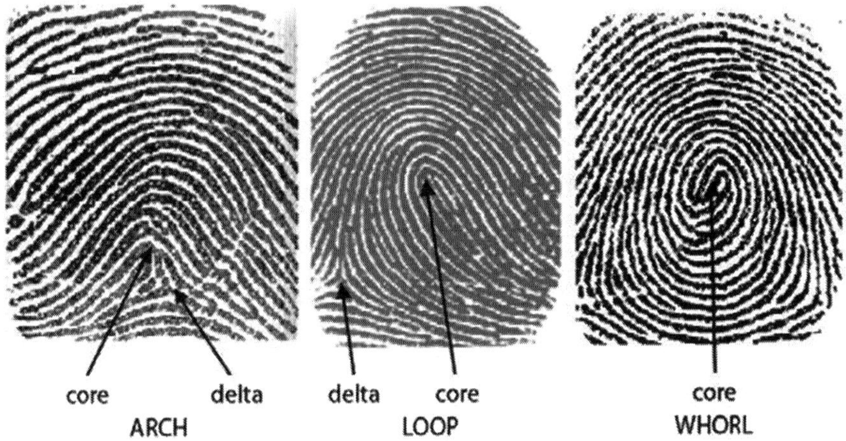

FIGURE 2.2 Fingerprint features
Source: Copyright 2008 L. O'Gorman

If two fingerprints are found to have identical characteristics, then it is likely that they belong to the same person's finger. Fingerprints are suitable for identification purposes due to their uniqueness, constancy throughout life and the fact that the patterns formed are suitable for systematic classification (Jackson & Jackson, 2008; Saferstein, 2015). However, it should also be noted that approximately four per cent of the population have fingerprints that cannot be effectively used for biometric identification purposes. This is caused by a range of factors, most commonly through injuries, such as burns, or as a result of carrying out manual labour over many years. For this reason it may not be possible for fingerprint recognition to be used in a system designed to have universal coverage of an entire population (Dessimoz & Champod, 2006). Often, such systems will combine more than one form of biometric to address this issue.

In addition to the unique nature of fingerprints, their easy accessibility and non-intrusive nature, and their cost effectiveness, have made them the most widely used contemporary form of biometric identification. Fingerprint scanning accounts for at least half of the biometric market worldwide. As will be discussed later, it has been adopted by a large number of private sector organisations for verifying the identity of employees, as well as by governments, including the United Kingdom and the United States, primarily for identifying suspects in criminal investigations, or travellers entering and leaving the country (NSTC, 2006).

Historically, fundamental work on the classification of fingerprints, as the process is understood today, is most closely associated with Sir Francis Galton (1882–1911). Galton was a British scientist and anthropologist, although others also made important contributions to its development. The systematic organisation of thousands of individual fingerprints into a searchable database on the basis of their characteristic architecture was undertaken by Sir Edward Henry (1850–1931). The Henry classification system provides a method to classify fingerprints and exclude potential match candidates. The work of Galton and Henry was a major impetus in the development of the technique of human fingerprinting as a basis for individual identification. This work has provided the foundation for the eventual establishment of fingerprint databases in law enforcement agencies around the world in the twentieth century (Allen, Sankar & Prabhakar, 2005).

The use of fingerprints for human identification purposes developed significantly following technological advancements that enabled computer technology to digitally retrieve and match relevant fingerprint data. Prior to this development in the latter part of the twentieth century, the only method available to law enforcement agencies to store, search and retrieve individual fingerprints was through a manual, card-based system. At that point in time, in order to match a fingerprint it was necessary to undertake a manual comparison of cards that were part of a large collection. This process was burdensome and time-consuming. As will be discussed later in this chapter, advancements in information technology have facilitated the creation of large, automated fingerprint databases that only require human input at the final stage, to distinguish between highly similar fingerprints as part of a list of close matches (Moses et al., 2010).

Contemporary biometric fingerprint recognition compares the unique combination of patterns on individuals' fingerprints through an automated digital process. An optical scanner is used to scan the finger. During this process, a laser illuminates the fingerprint ridges and converts the resulting image into a digital format. An algorithm filters out distortions caused by factors such as sweat or dirt, and further enhances the definition of the ridges in the image. In most digital systems in use today, an image of approximately 50 to 250 minutiae is obtained from the fingerprint by the scanner; but between 10 and 100 minutiae are actually used by the algorithm in creating the digital template of the fingerprint (Dessimoz & Champod, 2006).

The concept of fingerprint identification has been widely accepted for over a century and, in that time, has become established as one of the primary means of personal identification. More recently, the digitalisation of fingerprints and the efficiency offered by automated biometric technology, an attractive prospect for law enforcement and border security agencies around the world, has been widely adopted. National agencies in the United States and the United Kingdom led the world in funding research, sharing resources and collaborating with private industry to develop automated fingerprint identification systems capable of including hundreds of millions of fingerprints (Ashbourn, 2014).

Databases

Automated fingerprint matching databases were developed in the late 1990s; and today are known as Automated Fingerprint Identification Systems (AFIS). AFIS initially require that an optical device scans and uploads digital images of fingerprints to a centralised database. The distinctive architecture of the fingerprints is analysed to create a digital template representing key points in the fingerprint. A database operator can search the system to determine the correlation between two or more fingerprints, based on scoring criteria that they nominate. The system automatically produces a list of fingerprints stored in the database that have the closest match. Following this, a human fingerprint expert with many years of training in the field makes the final determination on whether the fingerprints match and belong to the same individual (Milne, 2013).

Automated databases used by law enforcement are typically comprised of two subsystems: a ten-print criminal identification system comprising a set of fingerprints obtained through an arrest or during the course of an investigation; and prints that are on file comprising latent fingerprints that have been obtained from crime scenes or physical evidence (Moses et al., 2010).

These subsystems enable fingerprint databases to conduct the following four types of searches:

- print-to-print searches: these are conducted to verify the identity of a suspect through a comparison of fingerprints obtained from a suspect against fingerprints stored in the database;
- mark-to-print searches: these are conducted in order to compare a fingerprint obtained from a crime scene, or other physical evidence, against fingerprints held within the database;
- print-to-mark searches: these are used to determine whether an individual is linked to other crime scenes by comparing their fingerprints against all the fingerprints held within the database, but in instances where previous searches have failed to produce a match;
- mark-to-mark searches: used to determine if a fingerprint obtained from a crime scene or physical evidence is connected with other prints held within the database.

(Moses et al., 2010)

AFIS have now been established in many jurisdictions around the world, and biometric fingerprint identification continues to be a primary method of establishing identity for law enforcement and border protection agencies. The next section of this chapter will describe some of the most prominent AFIS in more detail, providing more context and examples of contemporary fingerprint identification databases (Table 2.1).

In the United States, the Integrated Automated Fingerprint Identification System (IAFIS) is the national fingerprint database that has been operated by the Federal Bureau of Investigation (FBI) since July 1999. The IAFIS provides digital storage, retrieval and exchange capabilities for fingerprint images, and has an automated search capability. In addition to fingerprints, the database also includes biographical information in association with an individual's fingerprints, for example, information relating to their residential address, social security data and criminal history.

The IAFIS provides services to the FBI and security agencies, as well as state and local law enforcement agencies throughout the United States. Because the IAFIS provides shared access and integration across various law enforcement agencies, it plays an important role in the exchange of a range of other information, in

TABLE 2.1 Comparison of national fingerprint databases in the United States, United Kingdom and Australia

Jurisdiction	*Managing agency*	*System*	*Details*
United States	Federal Bureau of Investigation (Criminal Justice Information Services Division)	Integrated Automated Fingerprint Identification System (IAFIS)	IAFIS provides the capacity for electronic storage, retrieval and exchange of fingerprint images; as well as an automated search capability for the FBI and the federal, state and local law enforcement agencies within the United States, and includes criminal histories. IAFIS is currently being developed into the Next Generation Identification (NGI) system. This is a multimodal identification platform that includes fingerprints, facial photographs, palm prints, iris recognition data and an activity notification system.
United Kingdom	Partnership between private sector (Northrop Grumman) and the UK Home Office	National Automated Fingerprint Identification Service (IDENT1)	IDENT1 was developed in partnership with the private company Northrop Grumman, and superseded the National Automated Fingerprint Identification System (NAFIS-UK). It provides an integrated national fingerprint and palm-print system to police forces throughout the United Kingdom, and is linked to the Police National Computer, allowing for the integration of other biometric modalities as they become available.
Australia	Australian Criminal Intelligence Commission	National Automated Fingerprint Identification System (NAFIS)	NAFIS provides a centralised national database for electronic fingerprints and palm prints, an automated search and matching capability and national standards for each of Australia's policing and security agencies. As is the case with the other jurisdictions discussed, the system is currently in the process of being merged into a larger, multimodal biometric data system.

addition to identification, this includes information about persons of interest, suspects, offender profiles, criminal history and intelligence relevant to investigations and operations across many jurisdictions throughout the country (FBI, 2017).

Law enforcement and security agencies in the United States that have access to IAFIS can upload digital fingerprints to the database for matching, and are provided with a list of potential candidates for final verification by a human analyst. If a search of the database fails to produce any potential matches, the user is able to upload the fingerprints to the unsolved latent file sub-database. These fingerprints are automatically cross-referenced against the database when all new non-identical fingerprints are added, and the user is notified if, or when, a match can be established in the future. IAFIS has the capacity to include non-electronic fingerprint data from law enforcement agencies that do not have the capacity to submit electronic fingerprints. This capability is undertaken through the card checking service, which converts paper-based fingerprint information into a digital format (Moses et al., 2010).

IAFIS is in the process of being replaced by a new biometric identification service, known as the Next Generation Identification (NGI) system. The main difference between the existing system and the NGI system, is that the latter provides a platform for a multimodal functionality, adding several stages of advanced capability. These include the Advanced Fingerprint Identification Technology, which replaces the AFIS component of the system and will provide enhanced processing speed and accuracy. A repository for individuals of special concern has also been added, providing law enforcement agencies with the ability to conduct information searches, for example, relating to intelligence and data on wanted persons, sex offenders, terrorists and other individuals of special interest. Further, biometric capabilities that are already being added include the integration of palm prints, iris recognition, and facial recognition, to enhance the accuracy and scope of the system (FBI, 2017).

In the United Kingdom, the national automated fingerprint identification system is known as IDENT1. Unlike the other national fingerprint databases discussed, IDENT1 was developed as a joint venture between the research and development branch of the Home Office, and a private sector contractor. Since 2004, IDENT 1 has been developed, implemented and maintained by Northrop Grumman, a defence technology company headquartered in the United States. The new IDENT1 system has expanded the capability of the previous database to add a palm-print search and matching capability. Prior to that time, individual police forces in the United Kingdom had been electronically collecting and storing palm-print information; however, there was no national search capability. The IDENT1 computing infrastructure comprises over 1000 workstations and approximately 500 fingerprint scanning units. The system provides a link between 57 law enforcement agencies in England, Wales and Scotland, has a dedicated data communications network and is integrated with criminal records held on the Police National Computer (Northrop Grumman, 2017).

The Australian fingerprint database has operated since 2001 and is known as the National Automated Fingerprint Identification System (NAFIS). It is operated by the Australian Criminal Intelligence Commission (ACIC) following the merger of the Australian Crime Commission and the CrimTrac Agency in 2016. The NAFIS provides Australian law enforcement agencies, including the Department of Immigration and Border Protection, with a centralised national database for finger and palm print images, including an automated search and matching capability. The database establishes national standards for the collection, storage and searching of fingerprint and palm-print information held by Australian police, security and immigration agencies (ACIC, 2017).

The NAFIS enables real-time fingerprint uploads from individuals and crime scenes. When a person is processed by an Australian law enforcement agency, a digital image of their finger and palm prints is taken and uploaded to the NAFIS in order to verify their identity, or search against unidentified fingerprints to potentially link the individual with unsolved offences and crime scenes. Similarly, fingerprint evidence obtained from a crime scene can be uploaded to the NAFIS in order to compare it against existing crime scene images for intelligence purposes (ACIC, 2017).

The Biometric Identification System (BIS) is scheduled to replace the NAFIS in 2018. It is expected that the new system will improve police agencies' access to fingerprint data and provide integration with other forms of biometrics, including facial images. In response to concerns about privacy issues and the significant amount of money and resources being invested in biometric data, the Government highlights an increasingly complex security environment:

> Modern day threats demand agile IT capability that delivers greater and quicker collection of evidence, which can then be accessed nationwide. This is vital in the current national security landscape, because it is essential to have robust and efficient cross-border information sharing to support the law enforcement agencies that protect our communities. It's also vital our authorities are one step ahead of the sophistication of organised criminal syndicates who are adopting new and advanced technologies to exploit Australians.
>
> *(Australian Government, 2016)*

Applications and issues

Border security and international data sharing

In recent years, the implementation of e-passports and data systems with the capacity to automatically verify travellers' identity using biometric information has reduced the requirement for officials to conduct a manual passport and identity check. It is now possible for an automated process to compare the data stored in the electronic passport document with a live sample captured while the subject is crossing the border (Labati et al., 2015).

It has been increasingly common for developed countries to integrate biometric fingerprint identification (along with facial recognition) into their immigration systems since the mid-2000s. The requirement that foreign nationals and visa applicants submit biometric fingerprint data was introduced by the United States in 2004, Japan in 2007 and the United Kingdom in 2008. The European Union began collecting biometric fingerprints in October 2011, and Canada has required that some categories of foreign nationals provide biometric fingerprints prior to entering the country since 2013 (Canadian Government, 2017).

In Australia, fingerprint scans have been taken from individuals in immigration detention since December 2007, and collected from eligible protection visa applicants since 2009. This has been particularly beneficial for visa applicants who do not have identity documents:

> It may also stop people taking someone else's identity or nationality. Protection visa applicants are not always able to provide documentary evidence of their identity and/or nationality. This may be due to circumstances such as the applicants fleeing from persecution in their home country, their documents being destroyed in conflict, or arriving on fraudulent documents.
>
> *(Australian Government Department of Immigration and Border Protection, 2017)*

Anyone who makes a visa application to travel from designated countries must provide fingerprints and a photograph as part of their visa application. There are more than 20 designated countries across Europe, Africa, the Middle East and Asia. Collection of biometric fingerprints from passengers in airports began in Australia in early 2013. Current legislation allows biometric fingerprint scans to be conducted on any person in immigration clearance, such as passengers who are interviewed, to assist with their identification. Under section 257A of the *Migration Act 1958* (Cth), hand-held devices are being used to conduct fingerprint scanning of all international travellers at Australian airports typically taking less than 60 seconds to complete. According to the Department, fingerprint scans will not be retained: 'The scans will be deleted as soon as the check is completed' (Australian Government Department of Immigration and Border Protection, 2017a).

In recent years, Australia has introduced biometric fingerprint identification into its immigration system to complement the facial recognition technology that is already established. Legislation introduced in Australia in 2014, as part of a tranche of national security legislation, such as the *Counter-Terrorism Legislation Amendment (Foreign Fighters) Act 2014* (Cth) and the *Migration Amendment (Strengthening Biometrics Integrity) Act 2015* (Cth) facilitated the collection of biometrics, including facial and fingerprint data, from citizens and non-citizens entering or leaving the country for use in an automated border clearance system, known as a 'SmartGate' (Australian Government Department of Immigration and Border Protection, 2017b). This has become well established despite criticism regarding the breadth and scope of the discretionary power to collect biometric information.

International developments in data sharing over the past ten years provides further evidence of the ever-increasing collection of biometric data. For example, in 2009, a Five Country Conference (FCC) on biometric information sharing was established between Canada, the United States, the United Kingdom, Australia and New Zealand. A *High Value Data Sharing Protocol* between these countries has since allowed for the exchange of biometric fingerprint records to detect identity fraud and enhance security screening (Canadian Government, 2017). However, under this programme, shared fingerprints are not retained indefinitely. The New Zealand Government states that: 'When one member asks for fingerprint information from another, the receiving country destroys the fingerprint if it finds no match. Members do not share biographical data like names or personal details unless they match a fingerprint' (New Zealand Government, 2017). The Australian Government has also described the data sharing agreement:

> To combat identity fraud, we have entered into international information sharing agreements with a number of countries, including but not limited to, the United States, the United Kingdom, Canada and New Zealand. These international information exchanges may involve the sharing of personal identifiers, including facial images and fingerprint data. Where there is a match, additional information may be shared, such as biographical data, copies of travel and other identity documents or information from such documents, immigration status and immigration history and any criminal history information relevant to immigration purposes. Exchanged information can be checked against records in Australia. There are plans to increase the volume of data shared and to extend information exchanges and fingerprint match programmes to other countries.
>
> *(Australian Government Department of Immigration and Border Protection, 2017c)*

The Office of the United Nations High Commissioner for Refugees (UNHCR) has developed a biometric identity management system to improve the identification of refugees involved in its aid programs. In fact, the UNHCR has recorded fingerprint biometrics of tens of thousands of refugees, supplemented with iris recognition to enhance the level of accuracy they are able to attain. Also of note is their requirement for biometric technology to be portable and suitable for use in remote regions with limited infrastructure. The system that has been adopted utilises a portable Universal Serial Bus (USB) driven scanner that can be operated with limited power requirements and does not require Internet access (Lodinová, 2016).

Impact on criminal investigations

A number of empirical studies are available providing evidence of the impact of biometric fingerprint identification on criminal investigations. However, a large proportion of the literature focuses on descriptions of the technique, forensic investigation techniques and technical evaluations relating to independent systems or matching algorithms.

In the United States, a study examining the effectiveness of Minnesota's AFIS system demonstrated improvement in the detection and identification of offenders as technology improved over time, enabling investigators to obtain higher-quality prints from crime scenes to obtain more matches. At the time the research was conducted in the mid-2000s there was an average match rate of 20 per cent, in comparison to equivalent research conducted in the early 1990s with Kentucky's AFIS system, in which average match rates of approximately 3 per cent were calculated (Bradbury & Feist, 2005; Cordner, 1990).

An evaluation of the national fingerprint database in the United Kingdom was published in 2004 analysing how police in five jurisdictions obtained fingerprint evidence in cases of volume crime, such as burglary and motor vehicle thefts. The research demonstrated that the system provided a greater capacity to identify suspects and allowed for greater speed and efficiency in reaching case outcomes than would otherwise be possible, with investigators able to identify possible suspects more efficiently (MHB, 2004; Saferstein, 2015).

The introduction of the national fingerprint database in the United Kingdom led to changes in police practice, including the introduction of an expedited approach that prioritised the search for fingerprints at the scenes of volume crimes to increase the likelihood of recovering stolen property. One study examined whether the availability of evidence from the national fingerprint database influenced the decision of the Crown Prosecution Service to take a case to court. Although the prosecution service was of the view that fingerprint identification is more relevant at the investigation stage than the prosecution stage; they did consider that the national fingerprint database contributed significantly to an increased number of older cases being brought forward for prosecution, due to older latent data being cross referenced when new crime scene prints became available, and this contributed to an increase in the total number of volume crime cases brought before the court (MHB, 2004). The increasing availability of high-quality fingerprint evidence derived from national fingerprint databases can potentially further improve the progress of criminal investigations and prosecutions, reduce the number of contested cases and increase the likelihood that an offender would be found guilty.

The digitisation of fingerprint identification through automated databases has led to a significant number of positive identifications and linkages between individuals and physical evidence at other crime scenes, facilitating more targeted investigations. In particular, there is evidence that national fingerprint databases result in an increased likelihood of positive identification and linkage. In Australia, in the 2007–2008 financial year, there were 298,790 searches for fingerprints on the national database, which resulted in 31,219 identifications. In 2013–2014, 420,188 searches were conducted, resulting in 60,398 identifications, representing an increase in the rate of matches over this period from 7.4 per cent to 15 per cent (CrimTrac, 2014).

Acceptance and private sector applications

Fingerprints are the most established and accepted method of biometric identification in use today. While there may be a historical association between fingerprinting and criminality, its use continues to be applied in new contexts and keeps increasing. A recent survey conducted among members of the Biometrics Institute found that biometric fingerprint recognition was considered the most likely to dominate the field of biometrics over the coming years at 27 per cent, followed by facial (24 per cent) and voice recognition (7 per cent) (Biometrics Institute, 2015).

Sensor transducers for fingerprints are inexpensive to manufacture and can easily be integrated into existing electronics, such as mobile devices and security systems. Fingerprint recognition technology has wide application throughout the public and private sectors, and the fact that it is cost-effective means that it is the first form of biometric identification to be widely used in consumer electronic products at this point in time. It is currently generally available in mobile phones and laptop computers, as well as in building security systems. Continued expansion in the use of fingerprint identification in smartphones, laptops and tablets is likely (Ashbourn, 2014). Along with personal information technology products, its use in the banking sector, where security is a central aspect of the business, is also commonplace. This includes access to automated teller machines, e-commerce applications, protection of sensitive data held in computers and computer systems and for controlling access to restricted areas.

It is now common for mobile phones to have fingerprint recognition capabilities. These provide owners with the option of adopting a biometric login to unlock their devices by scanning their fingerprints rather than entering a password (Phonegg Global, 2017). The introduction of fingerprint scanning capabilities into mobile phones has provided an opportunity for the use of fingerprint identification in a range of commercial applications. It is now common for personal banking to be conducted online with biometric fingerprint identification of the customer via their mobile phone. Banks have been one of the first to introduce biometric fingerprint login options for online services using fingerprints instead of passwords (Head, 2014).

Commercial reasons direct the introduction of biometric identification systems, and these are often closely related to users' willingness to accept a new form of technology that integrates their physical characteristics. This is a factor that can determine whether or not a specific biometric identification system is used. For example, some users may be reluctant to use biometric fingerprinting because of a perceived association with criminality and law enforcement investigations.

There appears to be a different perception among consumers of the use of biometrics by government in comparison with the private sector. A 2014 survey found that 75 per cent of respondents were willing to submit biometric data to a government system through a fingerprint scan to confirm their identity at an airport gate when boarding a flight. However, in contrast, only 33 per cent of respondents were willing to have their biometric data used for retail offers at an airport (Unisys 2014).

A survey published in 2016 examined past and future use of biometric technology among individuals who had previously been victims of identity crime. It evaluated these individuals' previous use of biometric security mechanisms, and their willingness to use them in the future. Fingerprint identification was considered the most acceptable form of biometric identification to the survey participants. In the context of increasing awareness of security considerations in western countries as well as the potential impact of new technology on privacy rights, the study sought to examine whether there is a willingness to accept new technology as a security solution to inform future policy implementation by government (Tables 2.2 and 2.3).

The study found that among victims of identity crime the use of passwords is widespread, but only a small number had used biometric technology. Out of this sample, 96 per cent indicated that they would be prepared to use biometrics as an enhanced security measure, noting fingerprint recognition as the most acceptable, at 61 per cent; with approximately 30 per cent of respondents willing to use facial, iris or voice recognition (Emami, Brown & Smith, 2016).

A wide range of interesting new applications in fingerprint identification biometric technology are being commercialised and integrated into existing product lines by private companies, in order to increase their security or safety. Although this is more common with information technology products, there is also a trend towards introducing the technology to established non-digital product lines.

An example of an established non-digital product into which biometric fingerprint identification has been introduced is firearms. Known as 'smart guns' this type of firearms can be designed so that they can only be used by the registered owner. This can prevent the firearm from being used in a dangerous manner by someone other than the registered owner, such as a child, or prevent it from being sold onto the black market and used to commit a crime. In some cases, these firearms

TABLE 2.2 Previous use of security technologies

Technology	*n*	%
Passwords	394	88
Fingerprint	75	17
Facial recognition	30	7
Iris recognition	26	6
Voice recognition	25	6
Any	423	95

Source: Emami, Brown & Smith, 2013

TABLE 2.3 Willingness to use biometric technologies in future

Technology	*n*	%
Passwords	328	74
Fingerprint	270	61
Iris recognition	182	41
Facial recognition	164	37
Voice recognition	139	31
Any	427	96

Source: Emami, Brown & Smith, 2013

recognise associated biometrics such as hand size and grip technique (Simonetti, Rowhani-Rahbar, & Rivara, 2017).

In 2016, the US Government provided details of its intention to deploy personalised firearms technology to state employees, such as police, that incorporates biometric fingerprint identification in order for the firearm to be discharged. The technology is viewed in the United States as an opportunity to improve firearm safety, as part of a broader suite of reforms considered necessary to reduce well-documented societal harms caused by firearms (Simonetti et al., 2017). The wider roll-out of biometric fingerprint identification technology in firearms has been opposed by organisations such as the National Rifle Association (NRA). It has criticised the technology on a number of grounds, including that the time taken for the identification to take place may reduce the time available to the owner to respond in a life or death situation. The NRA is particularly opposed to a mandatory introduction of this form of security, that is, 'any law prohibiting Americans from acquiring or possessing firearms that don't possess "smart" gun technology' (NRA, 2017).

References

Allen, R., Sankar, P. & Prabhakar, S. (2005). Fingerprint identification technology. In *Biometric Systems: Technology, Design and Performance Evaluation*. London: Springer.

Ashbourn, J. (2014). *Biometrics in the New World: The Cloud, Mobile Technology and Pervasive Identity*. London: Springer-Verlag.

Australian Criminal Intelligence Commission (ACIC). (2017). *National Automated Fingerprint Identification System*. Retrieved from https://www.acic.gov.au/our-services/biometric-matching/national-automated-fingerprint-identification-system

Australian Government. (2016). *New National Biometrics System to Fight Crime*. Retrieved from https://www.ministerjustice.gov.au/Mediareleases/Pages/2016/SecondQuarter/New-national-biometrics-system-to-fight-crime.aspx

Australian Government Department of Immigration and Border Protection. (2017a). *Biometric Initiatives*. Retrieved from https://www.border.gov.au/about/corporate/information/fact-sheets/84biometric

Australian Government Department of Immigration and Border Protection. (2017b). *Your Personal Identifying Information*. Retrieved from https://www.border.gov.au/Forms/Documents/1243i.pdf

Biometrics Institute. (2015). *Biometrics Institute Industry Survey 2015*. Sydney: Biometrics Institute. Retrieved from http://www.biometricsinstitute.org/pages/industry-survey.html

Bradbury, S. & Feist, A. (2005). The use of forensic science in volume crime investigations: A review of the research literature. *Home Office Report 43/05*. Retrieved from https://www.gov.uk/government/uploads/system/uploads/attachment_data/file/115849/hoor4305.pdf

Canadian Government. (2017). *International Use of Biometrics*. Retrieved from http://www.cic.gc.ca/english/department/biometrics-international.asp

Cordner, G. (1990). *Estimating the Impact of Automatic Fingerprint Identification in Kentucky*. Frankfort: Kentucky Criminal Justice Statistical Analysis Centre.

CrimTrac (2014). *CrimTrac Annual Report 2013–2014*. Canberra: CrimTrac. Retrieved from https://www.acic.gov.au/files/crimtrac-annual-report-2013-14

Dessimoz, D. & Champod, C. (2006). Linkages between biometrics and forensic science. In *Handbook of Biometrics* (pp. 425–459). Boston, MA: Springer.

Emami, C., Brown, R. & Smith, R. (2016). Use and acceptance of biometric technologies among victims of identity crime and misuse in Australia. *Trends and Issues in Crime and Criminal Justice No. 511*. 1, Canberra: Australian Institute of Criminology.

Federal Bureau of Investigation (FBI). (2017). *Integrated Automated Fingerprint Identification System*. Retrieved from http://www.fbi.gov/about-us/cjis/fingerprints_biometrics/iafis/iafis

Head, B. (2014). Westpac launches fingerprinting access for online banking on iOs and Android devices. *The Age*. Retrieved from http://www.theage.com.au/it-pro/business-it/westpac-launches-fingerprinting-access-for-online-banking-on-ios-and-android-devices-20141205-120yr9.html

Irish Council for Bioethics. (2009). *Ethically Speaking*. Retrieved from https://repository.library.georgetown.edu/handle/10822/1026098

Jackson, A. & Jackson, J. (2008). *Forensic Science* (2nd edition), Harlow: Pearson Education.

Jain, A., Ross, A., & Prabhakar, S. (2004). An introduction to biometric recognition. *IEEE Transactions on Circuits and Systems for Video Technology 14*(1), 4.

Labati, R., Genovese, A., Munoz, E., Piuri, V., Scotti, F. & Sforza, G. (2015). *Automated Border Control Systems: Biometric Challenges and Research Trends*. Heidelberg: Springer.

Lodinová, A. (2016). Application of biometrics as a means of refugee registration: Focusing on UNHCR's strategy. *Development, Environment and Foresight 2*(2), 91. Retrieved from http://www.def-journal.eu/index.php/def/article/view/34

MHB (Morgan Harris Burrows). (2004). The processing of fingerprint evidence after the introduction of the national automated fingerprint identification, Home Office Online Report 23 April, London: Home Office.

Milne, R. (2013). *Forensic Intelligence*. Boca Raton, FL: CRC Press.

Moses, K., Higgins, P., McCabe, M., Probhakar, S. & Swann, S. (2010). Automated fingerprint identification system. In *Fingerprint Sourcebook*. Washington, DC: National Institute of Justice.

National Rifle Association (NRA). (2017). *Smart Guns and Personalised Firearms*. Retrieved from https://www.nraila.org/issues/smart-gunspersonalized-firearms

National Science and Technology Council (NTSC). (2006). *Privacy and Biometrics: Building a Conceptual Foundation*. Retrieved from https://www.hsdl.org/?view&did=463913

New Zealand Government. (2017). *Biometric Information*. Retrieved from https://www.immigration.govt.nz/about-us/policy-and-law/identity-information-management/how-biometric-information-is-used

Northrop Grumman. (2017). *IDENT1 Automated Fingerprint System, United Kingdom*. Retrieved from http://www.homelandsecurity-technology.com/projects/ident1-automated-fingerprint-system-northrop-grumman-uk

O'Gorman, L. (1999). *Fingerprint Verification in Biometrics: Personal Identification in a Networked Society*, Boston, MA: Kluwer Academic Publishers.

PhoneggGlobal. (2017). *Cell Phones with Fingerprint Scanner*. Retrieved from http://www.phonegg.com/list/182-Cell-Phones-with-Fingerprint-Scanner

Saferstein, R. (2015). *Criminalistics: An Introduction to Forensic Science*. Harlow: Pearson Education.

Simonetti, J., Rowhani-Rahbar, A. & Rivara, F. (2017). The road ahead for personalized firearms. *JAMA Internal Medicine 177*(1), 9.

Unisys. (2014). *Unisys Security Index Report Australia: Biometrics in Airports*. Retrieved from http://www.unisyssecurityindex.com/system/resources/uploads/113/original/In%20what%20circumstances%20are%20Australians%20willing%20to%20use%20biometrics%20in%20airports%20-%20May%202014.pdf?1400743365

3

DNA IDENTIFICATION

Introduction

Deoxyribonucleic acid (DNA) followed fingerprinting as the second major scientific technique in human identification for law enforcement purposes. DNA identification has not historically been included in texts on biometric identification. One reason for this may be the more substantial time and resources needed to create a DNA profile and enrol a subject in a database: this process may take hours rather than being instantaneous. Unlike other forms of biometric identification, the analytic techniques used derive from the biological sciences rather than the physical sciences, as other biometric techniques such as fingerprint and facial recognition do. However, DNA is clearly a biometric indicator, and further, the historical learnings from its development within the criminal justice system are significant and contribute to understanding the other forms of biometrics discussed throughout this text.

This chapter examines DNA identification, a technique used for over 30 years in criminal investigations. The use of DNA identification by law enforcement agencies continues to expand as new techniques are developed and technology enables it to be applied more efficiently. The chapter begins with an examination of the scientific and historical background of DNA identification, followed by a discussion of DNA databases, more recent techniques in this field and the application of DNA in criminal prosecutions.

Scientific and historical development

DNA identification is arguably the most significant scientific advancement in the history of forensic science, and it regularly plays an important role in modern criminal investigations of serious crimes. Despite the fact that there is a strong scientific foundation underpinning DNA identification, its application in the legal system has, at times, been controversial. The comparison of DNA profiles obtained from a crime scene with those from a suspect or database, is widely used in cases involving serious crimes against the person, particularly homicide and sexual assault. A range of new techniques of DNA identification continue to be developed and applied in criminal investigations around the world (Smith, 2015).

DNA can be recovered from most biological material. The most common human biological materials submitted for testing are blood and semen; as well as hair, saliva, skin and sweat. It can be obtained by analysing material present on personal items such as razors, hairbrushes or toothbrushes. DNA evidence is used to link or exclude an individual from association with the crime scene (notwithstanding the potential for the evidence to have been planted, or other explanations). The sample collection must accord with standard procedure, and a chain of custody must be established to enable DNA evidence to be used at trial (Butler, 2005).

At a crime scene, the forensic scientist must identify whether a sufficient amount of biological material is present to enable a sample to be taken. Measures must be taken to ensure that the biological material is properly preserved and not contaminated: such as wearing latex gloves, a mask over the nose and mouth and a hair net. Samples must be stored in designated evidence bags and administrative details carefully noted, such as a case and item number, date and the collector details, to ensure it can be admitted as evidence at trial.

DNA evidence obtained at a crime scene is analysed and then compared with biological material collected from a suspect. A sample from a suspect is usually obtained by pressing a cotton tip against the inside of the suspect's cheek, which painlessly removes mucosal cells (known as a buccal swab). The sample is then taped to collection paper for preservation, and preserved in a cold, dry environment (Butler, 2005).

A key technique used in DNA identification is *polymerase chain reaction* (PCR), a method developed in the 1980s by Dr Kary Mullis (Mullis & Faloona, 1987). PCR enables the rapid replication of a DNA sequence, and has facilitated dramatic advancements in molecular biology and forensic science. PCR has enabled DNA identification to advance, both in terms of its power of discrimination, and in its ability to obtain information from very small amounts of biological material. It is an enzymatic process that amplifies specific regions of DNA through cycles of heating and cooling (Saki & Mullis, 1988).

The fact that the human genome is unique means that it can be used as a form of identification. Repetitive parts of DNA within the genome, called *short tandem repeats* (STRs), exhibit variation between individuals. A DNA profile is created by analysing the number of STRs that occur at specific points in an individual's DNA. The STRs used for DNA identification are present in *non-coding* regions of the human genome: these regions do not provide any health or other information about the individual beyond their identity. A match between a DNA profile from a crime scene sample and a DNA profile from a suspect sample provides strong support for the inference that the samples are from the same person. However, there are alternative hypotheses that can account for a match, such as sample contamination. DNA identification must therefore be considered in the context of the other evidence (Smith, 2015).

An example of a DNA profile is the following gender designation and set of 13 paired numbers:

XY 9,12 18,21 14,14 15,16 25,28 14,16 11,10 29,30 15,16 8,10 12,20 8,11 7,19

These numbers specify the STRs at 13 points in the human genome. The numbers are paired at each point because one STR is inherited from each of the individual's parents.

DNA identification was first used in a criminal investigation in 1987, in the United Kingdom. In a high profile case in which police had made little progress, Professor Alec Jeffreys, was asked to analyse biological samples recovered from the bodies of two girls who were murdered in Leicestershire, and compare them with a sample of a suspect who had confessed to raping and murdering one of the victims. However, DNA identification established that the suspect's DNA did not match the sample recovered from the victim, and he was released. A screening of a subset of the men from three surrounding villages was conducted, and although a match to the crime scene profile was not obtained, it emerged that one man, Colin Pitchfork, coerced another into providing a sample on his behalf. It was later found that Pitchfork's DNA profile matched one found at the crime scene and he was subsequently convicted (Jobling & Gill, 2004).

Although the primary application of DNA identification has been in criminal law, it has been successfully used in other applications. These include to establish identity in visa applications and paternity cases, and to identify soldiers killed at war. An online industry offering mail order paternity testing kits, and tests diagnosing genetic disorders, has recently emerged. Within the legal system, DNA identification has played an important part in establishing the innocence of convicted persons. For example, DNA evidence has been used to exonerate over 350 people in the United States, many of whom were on death row awaiting execution (Innocence Project, 2017).

The collection of samples by forensic investigators is a critical stage in the criminal investigation process. Although the scientific foundation of DNA identification is well established, there are a number of means by which human factors (intentional or deliberate) or error may compromise the validity of the results obtained in the laboratory. For instance:

> A suspect's DNA profile might match the profile found at a crime scene as a result of tampering with the crime scene or subsequent substitution of DNA samples. This might occur where the actual offender, a police investigator, or another person deliberately leaves a suspect's genetic sample at the crime scene. Alternatively, it is possible that a suspect's sample might later be substituted for the actual crime scene sample to falsely implicate the suspect in the offence.
>
> *(ALRC 96, 2003, p. 1095)*

The high profile trial of O.J. Simpson in the 1990s focused attention on the practices of crime scene investigators, and highlighted the consequences of errors when the evidence is presented in the courtroom. In the Simpson case, the large amount of television footage of the crime scene was used by the defence to demonstrate, for example, that investigators had entered the scene without protective clothing, had not worn protective gloves and had dropped swabs on the ground prior to placing them in collection bags. The defence asserted in court that these actions may have led to contamination and compromised the evidence (Edwards, 2005). This trial occurred in the early 1990s when forensic DNA identification was a new phenomenon, highlighting the fact that despite strong theoretical support for the evidence, if forensic collection procedures are not strictly followed at the crime scene, evidence can be significantly devalued at trial.

Key questions arising in DNA evidence cases include whether or not the evidence was lawfully obtained, whether it could have been planted at the crime scene, whether untested samples may be of significance to the case, whether the chain of custody was maintained and whether the suspect's samples could have been mislabelled or cross-contaminated during collection, storage or transportation (Smith & Mann, 2015). The implications of errors in the collection and analysis of forensic biological material at trial will be discussed in more detail later in this chapter.

In recent years, more creative avenues for collecting DNA evidence have begun to emerge. As will be discussed in more detail later in the chapter, there have been many documented cases of police covertly obtaining evidence in an attempt to enhance criminal investigations. For example, police have obtained evidence from eating and drinking implements, such as a cup or fork, after a suspect has eaten a meal in a fast-food restaurant; or after a suspect has been observed spitting on the pavement. The practice of covertly obtaining DNA evidence, prior to formally requesting a sample, appears to be occurring more frequently (Harmaon, 2008).

An offender was recently convicted of numerous counts of sexual assault and murder in a high profile case in the United States, following the covert collection of DNA evidence. Investigators followed the suspect into a pizza restaurant and collected DNA from eating utensils, establishing that the suspect's DNA matched the crime scene profile. The offender was convicted of ten murders and one case of attempted murder (Miller, 2010).

Further opportunities for obtaining DNA profiles include using blood obtained from mosquitoes or leeches to provide a DNA profile that can be used as evidence in a criminal trial. In an Australian case, police investigating a burglary found a leach at the crime scene. When they tested blood it contained, the DNA profile matched a profile on the database of a man who been arrested the previous year for drug possession, and he was subsequently convicted on the basis of this evidence (ABC, 2009).

The most significant case involving the retention of DNA evidence is *R v Marper & S* (2002) EWCA Civ 1275. This case concerned whether the *Criminal Justice and Police Act 2001* (UK) contravened Article 8 of the European Convention on Human Rights (the Convention). According to the facts, two individuals, including a 12-year-old child, were charged with separate offences. Biological samples were obtained, and their DNA profiles created and included in the National DNA Database (NDNAD). Following their acquittal, the South Yorkshire Police refused to destroy the biological samples and DNA profiles. The case was appealed to the House of Lords (*R v Marper & S* (2004) UKHL 39), followed by the European Court of Human Rights, which delivered its decision in December 2008 (*Case of S. and Marper v The United Kingdom* ECHR, 4 December 2008). The Court ruled in favour of Marper and S, finding that:

> the blanket and indiscriminate nature of the powers of retention of the fingerprints, cellular samples and DNA profiles of persons suspected but not convicted of offences, as applied in the case of the present applicants, fails to strike a fair balance between the competing public and private interests and that the respondent State has overstepped any acceptable margin of appreciation in this regard.
>
> *(para. 119)*

The case did not focus on whether police had the legal right to obtain the evidence, but whether the retention breached the right to private life of the individuals concerned, under Article 8 of the Convention, and the right to fair and equal treatment under Article 14 of the Convention. The case highlighted the apparently unfair distinction between individuals who had been suspected and charged with an offence but subsequently released without conviction; and those in the broader community who had never been suspected of committing, and never been charged with committing a criminal offence.

Following the *Marper* ruling in December 2008, the United Kingdom Government responded with a number of policy changes. The DNA profiles of children younger than 10 were removed from the database and legislative amendments were announced that complied with the Court's decision. Anyone convicted of a recordable offence will still have their DNA profiles retained indefinitely, however, under the amended legislation, the government committed to:

- destroying all original DNA samples, including mouth swabs, as soon as they are converted into a digital database profile;
- deleting the profiles of those arrested but not convicted of a serious violent or sexual crime after 12 years;
- deleting the profiles of anyone arrested but not convicted of other offences after six years;
- removing the profiles of young people arrested but not convicted, or convicted of less serious offences, when they turn 18.

(National DNA Ethics Group, 2009)

In the United States, the *Justice for All Act of 2004* allows DNA profiles to be included in the national database for anyone charged with an indictable offence, even if the charges are subsequently withdrawn. However, DNA profiles of arrestees who have not been charged may not be included, nor samples that are voluntarily submitted for the purpose of elimination from the scene of a crime. The *DNA Fingerprint Act of 2005* enabled an arrestee's profile to be uploaded to the database at the time of arrest; however, if the arrestee is not subsequently charged, the burden lies with the arrestee to file a court order stating that the charges have been dismissed (Privacy International, 2006).

The Fourth Amendment of the US Constitution governs the legitimacy of government intrusion into the lives of private citizens, protecting the 'right of the people to be secure in their persons … against unreasonable searches and seizures'. In order to be considered reasonable, a search needs to be supported by a warrant on the basis of 'probable cause': the reasonable belief that the individual has committed a crime. As will be discussed in the next part of the chapter, this supports the argument that the immediate suspicion cast on those who refuse to participate is unreasonable and reverses the presumption of innocence.

In the 2012 Supreme Court case, *Maryland v King* (567 U.S.), King was arrested on assault charges and his DNA profile collected and retained in the state DNA database. His DNA profile was later linked to an unsolved rape of which he was subsequently convicted. King argued that the DNA evidence should have been suppressed because the Maryland DNA collection legislation violated the Fourth Amendment. Although the Maryland Court of Appeals found the legislation was unconstitutional, and set aside the rape conviction, the Supreme Court overturned this decision and held that the collection and retention of DNA profiles in databases is a legitimate and constitutionally valid procedure to identify arrestees.

Other cases of interest from the United States include *Commonwealth v Cabral* (69 Mass.App.Ct. 68, 2007) where it was held that there is no violation of the Fourth Amendment when a police investigator, who is following a rape suspect, observes the suspect spit on the street, and collects the saliva (containing skin cells), and establishes a match with the sample recovered from a victim. Although the suspect did have a reasonable expectation of privacy in his saliva, when he expectorated on the street and did not retrieve it, he assumed the risk of the public witnessing the act and taking possession of the fluid. In *Cabral*, the court relied on *Commonwealth v Ewing* 67 (Mass.App.Ct. 531, 2006) which found no expectation of privacy in cigarette butts that had been disposed of following a police interview.

Police in most jurisdictions can obtain forensic biological material from volunteers who consent to provide a sample. If a volunteer is a suspect, they have the right to refuse, however even their behaviour in refusing the request can provide police with useful information about their guilt or innocence. In the first use of DNA identification in the 1980s the suspect was identified not through scientific analysis, but through his behaviour in response to a request that he provide a sample of biological material. Investigators in the Pitchfork case undertook a mass screening of the entire male population of Leicestershire, England. Pitchfork was apprehended after it became apparent that he had asked another man to provide a sample on his behalf, and his guilt was confirmed with a DNA match (Wambaugh,

1989). Observing the behaviour and body language of those asked to provide a biological sample can be as important to the investigation as the technology of DNA identification itself.

It is normally the case that the only association the majority of mass screening volunteers have with the crime is that they reside in the geographical area in which it was committed. It has been argued that requesting DNA evidence in mass screenings of a large number of volunteers potentially infringes a fundamental criminal law principle, the privilege against self-incrimination, 'because it forces individuals who are reluctant to undergo DNA profile surveillance to reveal that reluctance to investigators' (Gans, 2001).

Mass screenings are not frequently undertaken. One instance, in the year 2000 in the town of Wee Waa, Australia, involved the rape of an elderly woman, and police believed that the offender was a resident. Five hundred men in the town were asked to voluntarily attend an interview and provide a sample of biological material. During the course of interviewing the men and obtaining samples, the anxiety of one man was noted by police, and he confessed to the crime shortly after providing his sample (Smith, 2015).

In this context, an interesting issue has been raised by legal scholars, regarding the privilege against self-incrimination. This is the common law right of an individual not to provide material or answer questions which may tend to incriminate him or her in a criminal offence. An individual cannot be compelled to incriminate themselves and the evidence must either be freely volunteered or discovered by the police. In this case 'equating an individual's behaviour when asked to participate in DNA profile surveillance ... with that individual's consciousness of her or his own criminal guilt' (Gans, 2001). On this argument, mass screenings place the individual in a situation in which by refusing to participate they are providing information to the police. For suspects and offenders, most criminal procedure legislation provides the capacity for police to obtain biological material compulsorily. Requiring police to obtain a court order requesting that certain individuals submit samples would be a preferable approach, notwithstanding the significant public interest in prosecuting serious crimes

DNA databases

DNA identification databases refer to a collection of genetic sequence information that is used to identify specific individuals. This includes details of STRs, and potentially also phenotypic information. The following quote provides an example of how DNA databases are defined in criminal procedures legislation:

> a database (whether in computerised or other form and however described) containing (a) the following indexes of DNA profiles: a crime scene index, a missing persons index, an unknown deceased persons index, a serious offenders index, a volunteers index, a suspects index, and information that may be used to identify the person from whose forensic material each DNA profile was

derived; (b) a statistical index; and (c) any other index prescribed by the regulations.

(Crimes Act 1914 *(Cth), section 23YDAC, Australia)*

Large numbers of DNA profiles are collected and stored by law enforcement agencies to aid the investigation of serious crimes. The world's largest forensic DNA databases have been established in the United States and the United Kingdom. In January 2017, the US National DNA Index System (NDIS) contained over 12.5 million offender profiles and 2.6 million arrestee profiles (FBI, 2017). In March 2017, the UK National DNA Database (NDNAD) contained over 5.2 million individual profiles and over 500,000 crime scene sample profiles (Home Office, 2017). The Australian National Criminal Investigation DNA Database (NCIDD), currently managed by the Australian Criminal Intelligence Commission (ACIC), has been in operation since 2001, and holds more than 80,000 DNA profiles (ACIC, 2017). Several countries share DNA profiles internationally when relevant to investigations. For instance, in 2014, the Australian Government acknowledged that it had entered into a DNA profile sharing programme with the United Kingdom, the United States and Canada (Keenan, 2014).

Research has been conducted, providing empirical evidence of the contribution of DNA to criminal investigations, demonstrating that DNA identification does have a positive impact on criminal justice outcomes. A randomised study of the effect of DNA in property crime investigations across five locations in the United States used traditional investigative techniques in treatment and control groups, with the treatment group incorporating DNA identification. It was found that the use of DNA evidence in the investigation of property offences resulted in twice as many suspects identified, arrests and cases accepted for prosecution. A comparison between DNA and fingerprint identification was also undertaken, with the finding that DNA was five times more likely to result in the identification of a suspect, compared with fingerprint evidence (Roman et al., 2008).

Research conducted in Queensland, Australia, has examined the impact of DNA evidence on court outcomes for sexual offences, homicide and property offences. In these studies, half of the cases involved the presentation of DNA evidence in court, while the other half were assigned as comparison-control cases. In sexual offence cases, DNA evidence doubled the likelihood that a case reached court, and the presentation of DNA evidence by the prosecution at trial resulted in a 33-fold increase in the likelihood that a jury would find the offender guilty. DNA evidence also increased the likelihood of a custodial sentence. In relation to homicide cases, those where DNA evidence was presented by the prosecution were more than 14 times more likely to reach court, and juries were more than 23 times more likely to convict. Finally, for property crimes, the study indicated that cases with DNA evidence had an increased probability of reaching court, and the offenders were more likely to plead guilty (Briody, 2002, 2004, 2006). There appears to be strong empirical evidence supporting the use of DNA identification in criminal investigation, particularly with respect to property offences and serious offences against the person.

Criminal prosecution

DNA evidence is considered to be circumstantial evidence in a criminal case, insufficient for a conviction in itself, and normally only comprising part of the prosecution's case. For instance, if an accused has a strong alibi, it may be possible to be found not guilty despite the existence of DNA evidence. However, if there is further circumstantial evidence that demonstrates an association between an accused and a crime scene, DNA evidence is likely to be highly significant, as it is commonly a vital part of the prosecution's case, with other forms of circumstantial evidence playing a supporting role (Findlay & Grix, 2003). Despite the fact that DNA identification as a forensic technique has been established for over 30 years, it can be technically complex and the potential for jury misunderstanding may affect its probative value.

A match between a defendant's DNA profile and a crime scene sample is presented in court as a *match probability*. This refers to the probability that if another individual were selected from a population at random, they would have the same DNA profile – the same STR allele frequency at the points included in the profile. This can be potentially confusing in the context of a criminal trial if it is not explained and presented correctly. The prosecutor's fallacy involves a misrepresentation of the probative value of a match to the benefit of the prosecution, for example:

1. Only one person in a million will have a DNA profile which matches that of the crime stain.
2. The defendant has a DNA profile which matches the crime stain.
3. Ergo there is a million to one probability that the defendant left the crime stain and is guilty of the crime.

(Doheny and Adams v The Queen *(1997) 1 Cr. App. R. 369, 372–373)*

Although there is the potential for this fallacy to occur in relation to DNA evidence, the significance of the evidence depends on the other facts in the case. However, it was observed in a United Kingdom case that for a male defendant, 'DNA evidence tells us no more than the fact that there is a statistical probability that he was the criminal of 1 in 26' (*Doheny and Adams v The Queen* (1997) 1 Cr. App. R. 369, 372–273).

The empirical research discussed above indicates that the availability of DNA evidence increases the likelihood that a defendant will be convicted (Briody, 2004). Research findings that are based on interviewing jurors at the conclusion of actual trials or simulated trials raise questions about the capacity of jurors to understand and apply scientific evidence to facts presented in the courtroom (Findlay & Grix, 2003). Basic training on relevant scientific principles and interpretation issues for jurors, has been proposed to help ensure a sufficient level of understanding is present for effective decision-making (Wheate, 2008).

DNA identification has provided an opportunity to evaluate the accuracy of convictions and acquittals, and to examine the causative factors involved. This is the case with older cases prior to the 2000s when DNA technology first became widely used. Eyewitness error has been found to be the most prevalent factor contributing to wrongful conviction, with an incidence of approximately 60 per cent of all wrongful convictions identified (Huff, 2004). Errors associated with scientific evidence and false confessions obtained under duress are also considered to be significant contributors (Campbell & Denov, 2005). This may be due to investigators overlooking or suppressing evidence that supports a defendant's innocence due to pressure to gain a conviction, and is more likely in high profile cases where the accused is of low socio-economic status.

Innocence projects were first established in the 1990s when the potential of DNA evidence to overturn wrongful convictions became apparent. They have generated a great deal of publicity, and been involved in a number of high profile cases. The first innocence project was founded in 1992 at Cardozo Law School in New York. As of 2017, it has been responsible for 349 post-conviction DNA exonerations (217 of which involved African Americans). It provides legal representation, undertakes research into the causes of wrongful conviction, contributes to law reform efforts and seeks to raise awareness of wrongful conviction. The work of innocence projects has led to the identification of flaws in earlier forms of scientific technology. For example, convictions obtained in earlier cases using microscopic hair analysis were overturned when mitochondrial DNA identification could demonstrate that the hair samples presented at trial were not those of the person convicted of the crime (Innocence Project, 2017).

One of most prominent cases taken on by the Innocence Project in the United States involved Darryl Hunt. Hunt was convicted in 1984 of rape and murder in North Carolina. A witness told police that they had seen the victim with an African-American man who matched Hunt's description on the morning of the crime and identified him in a line-up. Although Hunt's girlfriend initially told police she was with him on the night of the crime, when he was later arrested on unrelated charges, she told police that he had confessed to raping the victim. Hunt was subsequently convicted and sentenced to life imprisonment. The conviction was overturned on appeal due to flaws in the evidence, but at a second trial, he was again convicted, supported by statements from prisoners that Hunt had confessed to the crime. In 2004, the DNA profile from semen found on the victim's body was checked against the state DNA database and found to match the profile of a prisoner serving a life sentence for murder. Darryl Hunt was subsequently exonerated after serving 18.5 years in prison, and awarded a significant settlement by North Carolina in 2007 (Innocence project, 2017).

In a 2009 Australian case, 22-year-old Farah Jama was released from prison after serving a 16-month prison sentence for raping a 48-year-old woman in a nightclub toilet. Despite the fact that DNA evidence was the only evidence that linked Jama to the crime, the fact that he had an alibi, and that there was no CCTV footage of his presence at the nightclub, he was nonetheless convicted. Contamination occurred at the forensic medicine centre when samples were taken from the victim in the same cubicle as a woman known to Jama who provided samples as part of an unrelated investigation. Following the incident, a review of all cases involving DNA evidence over the previous five years was announced, and the use of DNA evidence at trial suspended while a review was undertaken. In response to these events, public debate about the use of DNA identification evidence at trial recognised that there is a case for adopting a warning to juries not to place too much reliance on DNA evidence and the need to prevent similar errors being made in the future due to overreliance on DNA evidence (Vincent, 2010). More recently, in Western Australia in 2017, the leading forensic science expert in the state was sacked after systematically breaching protocols and casting doubt about the outcome of more than 27 convictions for serious crimes (Powell, 2017).

In the United Kingdom, a review body known as the Criminal Cases Review Commission (CCRC) has been established. The CCRC was created in 1995 as a result of the Runciman Commission, set up to determine whether the legal system was correctly convicting the guilty. The CCRC is an independent body established to conduct transparent, impartial and accountable investigations into suspected miscarriages of justice. It has the power to refer claims of wrongful conviction to the Court of Appeal in instances where there is a 'real possibility' that the conviction can be overturned on the basis of arguments not raised at trial, evidence not presented at trial or due to other exceptional circumstances. Between 1997 and 2017, the CCRC referred 629 cases for review, and of those 414 appeals against convictions were allowed (Select Committee on Home Affairs, 1999).

An issue that arose in the course of investigating cases of wrongful conviction, particularly those cases that involved DNA identification, was access to preserved samples. This has led to calls for specific DNA-based innocence testing legislation to be enacted. Legislation in Illinois and New York has served as a model in other parts of the United States for post-conviction DNA testing reform measures (Christian, 2001). The New York legislative model only applies to prisoners who were convicted prior to 1996. This represents the time that legislators believe DNA identification became sufficiently widely used, such that further DNA testing would not provide different results, and lead a court to reach a different outcome. The legislation requires that prisoners making an application demonstrate that it is reasonably probable that the outcome of their trial would have been more favourable if DNA evidence was used. The New York legislation states that:

> In cases of convictions occurring before January first, nineteen hundred ninety-six, where the defendant's motion requests the performance of a forensic DNA test on specified evidence, and upon the court's determination that any evidence containing deoxyribonucleic acid ('DNA') was secured in connection with the trial resulting in the judgment, the court shall grant the application for forensic DNA testing of such evidence upon its determination that if a DNA test had been conducted on such evidence, and if the results had been admitted in the trial resulting in the judgment, there exists a reasonable probability that the verdict would have been more favourable to the defendant.
>
> *(New York Criminal Procedure Law, section 440.30)*

In *People v Tookes* 639 N.Y.S.2d 915 (Sup. Ct. 1996), it was held that the 'reasonable probability' requirement referred to whether the DNA identification results would provide a 'reasonable potential for exculpation'. The Illinois legislation is also limited to prisoners who received their conviction prior to the availability of DNA testing technology:

> A defendant may make a motion before the trial court that entered the judgment of conviction in his or her case for the performance of fingerprint or forensic DNA testing on evidence that was secured in relation to the trial, which resulted in his or her conviction, but which was not subject to the testing which is now requested because the technology for the testing was not available at the time of trial. Reasonable notice of the motion shall be served upon the State.
>
> (Illinois Code of Criminal Procedure of 1963, *725 Ill Comp. Stat. 5/116–3)*

However, there are a number of further requirements stipulated by the Illinois legislation which are more stringent than the New York legislation. It must be established, *prima facie*, that the identity of the person responsible for the crime was at issue at trial, and that the DNA evidence that will be relied upon has been subject to a chain of custody. Further, the court must determine that the proposed DNA testing has the 'scientific potential' to create new evidence, and that it will be 'materially relevant' to the prisoner's assertion of innocence (Illinois Code of Criminal Procedure of 1963, 725 Ill Comp. Stat. 5/116–3).

It was held in *People v Gholston* (464 N.E.2d 1179, 1984) that to meet this standard, the presented evidence must be so conclusive, that 'it would probably change the result on a retrial'. This case related to a sexual assault involving multiple offenders. At trial, the Court held that post-conviction DNA testing would not offer the potential to exculpate the prisoner, because a comparison of his DNA profile and the profile obtained from semen deposited on the body of the victim, would not be able to exclude the possibility that he was one of the other offenders that were involved in committing the crime:

> In general, genetic testing of the type requested by the defendant has the potential to offer material evidence and, in some cases, evidence that could exonerate a defendant if no match is found. However, this is not true under the circumstances of the present case. Even if the defendant's DNA sample were found not to match the DNA taken from the victim, this result would not be material to his claim of actual innocence.
>
> (People v Gholston *(464 N.E.2d 1179, 1984), para. 13)*

Several barriers contribute to the amount of time that an investigation takes to complete, which may have implications for the retesting of DNA samples. In a number of jurisdictions, there is no requirement that crime scene samples be preserved, no legal right to obtain knowledge of their existence, no legal right to access the evidence and no legal right to have DNA testing performed for the purpose of establishing innocence. Law reform in this area has been developing slowly.

Individuals who have been convicted of a crime may seek access to crime scene samples for the purpose of reviewing their conviction. There are a number of reasons for pursuing this option. For example, they may have been convicted prior to the availability of the necessary technology; the prosecution may not have introduced relevant DNA evidence at trial; or the defence may not have sufficiently questioned the quality, probity or presentation of the DNA evidence at trial. Due to the fact that it is possible to preserve biological samples for decades, and because scientific progress is providing increasingly accurate analyses, an argument could be made that in some circumstances a more accurate outcome is possible over time, further supporting the value of retaining evidence indefinitely in key cases.

In the United States, a number of legislatures limit the post-conviction DNA testing of prisoners on death row, or those who have been sentenced to life imprisonment without parole. Post-conviction DNA testing may be a temporary matter of concern. Once DNA identification is conducted in all investigations where biological material is located, and as scientific results become more accurate, demand for post-conviction DNA testing may decline to the point that the legislation is no longer necessary. A point may be reached where DNA exonerations are reduced to a very small fraction of cases. The number of prisoners that can be assisted by further DNA testing is declining because DNA technology has been so widely used since the mid-1990s and has effectively excluded thousands of innocent suspects before a case against them reached trial.

Currently, in most Australian jurisdictions, biological samples are held on a long-term basis, and prisoners can apply to the Director of Public Prosecutions or the relevant police service for access. Although the Australian Law Reform Commission had recommended that samples be permanently retained, they acknowledged that this may not be necessary or achievable in practice (ALRC, 2003).

New techniques

New techniques in this field have been developed in the past decade that extend beyond traditional DNA profile matching. These new techniques increase the potential for DNA identification to be used in new ways, not only in the criminal justice system, but also more broadly. A key factor in this will be the development of portable DNA testing devices that significantly reduce the cost, time and size of the equipment required for DNA identification. They are an important development that will contribute to DNA identification being considered a primary biometric identifier in the way that fingerprints and iris scans are today. Devices that can be connected to a smartphone and provide analysis and a DNA profile within ten minutes have been developed and are expected to be mass produced and widely available within the next five years (Chin, 2017).

DNA phenotyping is a new technique applied by criminal investigators that uses *coding* regions of the genome. It has been less widely used than the standard technique and has been more controversial. This is because DNA phenotyping can determine whether an individual has specific genes that are relevant to physically identifiable features, such as hair and eye colour, height, ethnic background, facial features and also predisposition to specific psychological and other medical conditions (Kayser & Schneider, 2009).

DNA phenotyping is used in cases where investigators cannot establish a DNA match on a database, and provides for information to be obtained about a suspect where there is otherwise little evidence available. In most jurisdictions around the world the technique remains unregulated. From an ethical standpoint, there do not appear to be privacy concerns about externally visible traits such as eye and hair colour, however, these concerns may arise when DNA phenotyping is conducted that reveals medical conditions (Smith & Urbas, 2012).

In a 2011 case from the United Kingdom – one of the few publicised cases to have relied on phenotyping – *R v Delroy Grant*, a prolific offender committed numerous burglaries and rapes over a 20-year period, and implemented strategies to avoid detection, such as disconnecting the electricity in victims' houses and wearing a mask and gloves. In the absence of any other investigative leads, police used DNA phenotyping to determine the suspect's race, which proved critical to the investigation and enabled investigators to identify him (Kopec, 2014).

DNA profiles can demonstrate familial relationships on the basis of the number of shared STR markers between two or more profiles. For instance, a child would share half of their parent's STR markers, as half of a person's genetic code is received from each parent. Familial searching is a new application of DNA identification that is increasingly being used in criminal investigations (McCarthy, 2011). If a match cannot be established with any of the DNA profiles held on a database, partial matches may indicate familial relationships that police can then investigate using traditional lines of inquiry. For instance, whether a relative lives in the vicinity of the crime scene (Greely et al., 2006).

The English case *R v Harman* demonstrates how familial searching can provide police with new leads in an investigation. The case involved a lorry driver who was killed when a brick was thrown from a footbridge onto a motorway. DNA evidence obtained from the brick elicited a partial match that showed that the suspect who deposited their biological material on the brick had a brother who lived near the crime scene. This person was located, eventually confessed to the crime and was convicted of manslaughter (Greely et al., 2006).

The use of this technique in a civil law context is known as kinship matching. It has been used in circumstances such as a plane crash or natural disaster incident to identify victims, involving multiple unidentified bodies, or in isolated cases of unidentified bodies being discovered, and can reduce the time required for this process to be undertaken. Family members of the victims voluntarily submit samples of their DNA to investigators who then seek to establish partial matches with biological material found at the scene. Where a partial match is found, this can reduce the time required to identify bodies in natural disaster and missing person cases. Kinship matching was used to identify victims of the 2014 Malaysia Airlines flight disaster in the Ukraine (Netherlands Forensic Institute, 2014).

Another more recently developed form of DNA identification involves mitochondrial DNA (mtDNA). Mitochondria are cell structures responsible for energy production and have their own DNA, which is outside the cell nucleus. Significantly, there are several hundred mitochondria per cell, meaning that it is more likely to be present in degraded samples of DNA, such as those likely to be found in degraded skeletal remains. Another important point is that mtDNA is maternally inherited and identical between siblings and individuals who are maternally related.

mtDNA identification can be used to identify familial relatedness, and in the identification of degraded samples, such as skeletal remains, where there may be insufficient nuclear DNA to create a profile (Coble et al., 2004). mtDNA can be traced back maternally for generations and be used to trace geographic ancestry back to specific countries and ascertain a suspect's likely ethnic background (Kopec, 2014).

A review of the historical development and issues associated with DNA identification provides important examples of the legal system accommodating new scientific developments in biometric identification more broadly. The techniques reviewed in this chapter have made significant contributions to criminal investigations. Issues such as the weight given to DNA evidence in circumstantial cases, and sample contamination, impact on the validity of DNA identification. The development of DNA phenotyping, familial searching, mtDNA identification and other techniques will continue to expand over coming years and be used in new contexts. Further research that examines how these should be regulated will be important. Despite ongoing technological advancements, DNA identification must continue to be considered as only one part of the overall evidence in a case.

As technology develops over the next decade, and DNA profiles can be created increasingly efficiently, it is likely that DNA identification will be more commonly viewed as a biometric identifier and applied more broadly beyond the criminal justice system.

References

Australian Broadcasting Corporation (ABC) News. (2009, 20 October). *Leech identifies robber.* Retrieved from http://www.abc.net.au/news/2009-10-20/leech-identifies-robber/1109838

Australian Criminal Intelligence Commission (ACIC). (2017). *National Criminal Investigation DNA Database.* Retrieved from https://www.acic.gov.au/our-services/biometric-matching/national-criminal-investigation-dna-database

Australian Law Reform Commission (ALRC). (2003). *Essentially Yours: The Protection of Genetic Information in Australia.* Retrieved from http://www.alrc.gov.au/publications/report-96

Briody, M. (2002). The effects of DNA evidence on sexual offence cases in court. *Current Issues in Criminal Justice 14*(2), 159.

Briody, M. (2004). The effects of DNA evidence on homicide cases in court. *The Australian and New Zealand Journal of Criminology 37*(2), 231.

Briody, M. (2006). The effects of DNA evidence on property offences in court. *Current Issues in Criminal Justice 17*(3), 380.

Butler, J. (2005). *Forensic DNA Typing*. Burlington, MA: Elsevier.

Campbell, K. & Denov, M. (2005). Criminal injustice: Understanding the causes, effects and responses to wrongful conviction in Canada. *Journal of Contemporary Criminal Justice 21*,224.

Chin, M. (2017, 20 March). UCLA researchers make DNA detection portable, affordable using cellphones. Retrieved from http://newsroom.ucla.edu/releases/ucla-researchers-make-dna-detection-portable-affordable-using-cellphones

Christian, K. (2001). And the DNA shall set you free: Issues surrounding post-conviction DNA evidence and the pursuit of innocence. *Ohio State Law Journal 62*, 1195.

Coble, M., Just, R., O'Callaghan, J., Letmanyi, I., Peterson, C., Irwin, J., & Parsons, T. (2004). Single nucleotide polymorphisms over the entire MtDNA genome that increase the power of forensic testing in Caucasians. *International Journal of Legal Medicine 137*, 143.

Edwards, K. (2005). Ten things about DNA contamination that lawyers should know. *Criminal Law Journal 29*, 71.

Federal Bureau of Investigation (FBI). (2017). *NDIS Statistics*. Retrieved from https://www.fbi.gov/services/laboratory/biometric-analysis/codis/ndis-statistics

Findlay, M. & Grix, J. (2003). Challenging forensic evidence? Observations on the use of DNA in certain criminal trials. *Current Issues in Criminal Justice 14*, 272.

Gans, J. (2001). Something to hide: DNA, surveillance and self-incrimination. *Current Issues in Criminal Justice 13*, 168.

Greely, H., Riordan, D., Garrison, N., & Mountain, J. (2006). Family ties: The use of DNA databases to catch offenders' kin. *Journal of Law, Medicine & Ethics 34*, 248.

Harmaon, A. (2008, 3 April). Lawyers fight DNA samples gained on the sly. *New York Times*. Retrieved from http://www.nytimes.com/2008/04/03/science/03dna.html

Home Office of the United Kingdom. (2017). *National DNA Database Statistics*. Retrieved from https://www.gov.uk/government/statistics/national-dna-database-statistics

Huff, R. (2004). Wrongful convictions: The American experience. *Canadian Journal of Criminology and Criminal Justice*, 107.

Innocence Project (2017). Retrieved from https://www.innocenceproject.org

Jobling, M. & Gill, P. (2004). Encoded evidence: DNA in forensic analysis. *Nature Reviews Genetics 5*, 739.

Kayser, M. & Schneider, P. (2009). DNA-based prediction of human externally visible characteristics in forensics: Motivations, scientific challenges and ethical challenges. *Forensic Science International: Genetics 3*, 154.

Keenan, M. (2014, November 6). Minister signs international DNA exchange pilot with United Kingdom. Retrieved from http://www.ministerjustice.gov.au/Mediareleases/Pages/2014/FourthQuarter/6November2014-MinisterSignsInternationalDNAExchangePilotWithUnitedKingdom.aspx

Kopec, M. (2014). A new use of 'race': The evidence and ethics of forensic DNA ancestryprofiling. *Journal of Applied Philosophy 31*, 237.

McCarthy, M. (2011). Am I my brother's keeper? Familial DNA searches in the twenty-first century. *Notre Dame Law Review 86*, 381.

Miller, G. (2010). Familial DNA testing scores a win in serial killer case. *Science 329*, 262.

Mullis, K. & Faloona, F. (1987). Specific synthesis of DNA in vitro via a polymerase-catalyzed chain reaction. *Methods in Enzymology 155*, 335.

National DNA Ethics Group, United Kingdom. (2009). *Second Annual Report*. Retrieved from https://www.gov.uk/government/publications/ndnad-ethics-group-2nd-annual-report

Netherlands Forensic Institute. (2014). Current situation regarding the DNA analysis of MH17. Retrieved from http://www.forensicinstitute.nl/about_nfi/news/2014/current-situation-regarding-the-dna-analysis-mh17.aspx?cp=34&cs=578

Powell, G. (2017, 31 March). Leading DNA scientist sacked, 27 criminal convictions in doubt, WA Attorney-General says. Retrieved from http://www.abc.net.au/news/2017-03-31/sacking-of-was-leading-dna-scientist-27-criminal-cases-in-doubt/8403618

Privacy International. (2006). The United States and the development of DNA data banks. Retrieved from http://www.privacyinternational.org/article.shtml?cmd%5B347%5D=x-347-528471#_ftn25

Roman, J., Reid, S., Reid, J., Chalfin, A., Adams, W. & Knight, C. (2008). *The DNA Field Experiment: Cost-effectiveness Analysis of the Use of DNA in the Investigation of High-volume Crimes*. Washington, DC: Urban Institute.

Saki, R., & Mullis, K. (1988). Primer-directed enzymatic amplification of DNA with a thermostable DNA polymerase. *Science 239*, 487.

Scheck, B., Neufeld, P., & Dwyer, J. (2001). *Actual Innocence: When Justice Goes Wrong and How to Make It Right*. New York: Signet.

Select Committee on Home Affairs. (1999). *First Report: The Work of the Criminal Cases Review Commission*. Retrieved from https://www.publications.parliament.uk/pa/cm199899/cmselect/cmhaff/106/10602.htm

Smith, M. (2015). *DNA Evidence in the Australian Legal System*. Sydney: Lexis Nexis.

Smith, M. & Mann, M. (2015). Recent developments in DNA evidence. *Trends and Issues in Crime and Criminal Justice 506*, 1. Canberra: Australian Institute of Criminology.

Smith, M. & Urbas, G. (2012). Regulating new forms of forensic DNA profiling under Australian legislation: Familial matching and DNA phenotyping. *Australian Journal of Forensic Sciences 44*, 63.

Vincent, F. (2010). *Inquiry into the Circumstances that Led to the Conviction of Mr. Farah Abdulkadir Jama*. Melbourne: Department of Justice.

Wambaugh, J. (1989). *The Blooding*. London: Bantam.

Wheate, R. (2008). Australian forensic scientists: A view from the witness box. *Australian Journal of Forensic Sciences 40*, 123.

4

FACIAL RECOGNITION

Introduction

This chapter examines biometric facial recognition, one of the most rapidly developing methods of biometric identification for police and government agencies around the world. There are a number of features of facial recognition that differentiate it from the other biometrics considered in this text. These include its capacity for integration with other technologies, such as closed circuit television (CCTV), to facilitate covert tracking. This chapter commences with an examination of the scientific and historical background of facial recognition, including its origins in forensic facial mapping. The applications of facial recognition are then examined, including border control, public surveillance and the identification of unknown suspects. The development of facial recognition databases in several jurisdictions around the world is then explored, highlighting the fact that many facial recognition databases are drawn from established administrative databases, such as driver's licence photograph databases. The chapter concludes with an overview of some of the emerging issues that relate to the establishment of facial recognition databases, and the increasing use of facial recognition technology, including accuracy, and existing regulatory oversight and accountability mechanisms.

Scientific and historical background

Since the nineteenth century, police have used photographs and artist sketches for the purposes of identifying unknown suspects. Facial comparison (often referred to as facial mapping) can be traced back to sketches of suspects in criminal investigations, made by portrait artists on the basis of witness statements (Valentine & Davis, 2016). Facial mapping involves the review of photographic and CCTV images by an expert, where the prosecution seeks to prove that the defendant is the individual depicted. Facial mapping procedures can involve either a quantitative method, where measurements between facial features are compared (photo-anthropometry), or a qualitative method that examines the similarities of facial features (morphological analysis) (Edmond et al., 2009).

Facial recognition is an extension of facial mapping, however, in contrast to this earlier method, an algorithm is used to position, extract, digitise and compare the arrangement of facial features (Ricanek, 2014). The algorithm used for facial recognition is similar to that used for fingerprint recognition (Adler & Schuckers, 2007). It enables the comparison of photographs of facial images with those stored on databases or identification documents. The process has a number of stages, commencing with a digital photograph being taken and analysed via face detection methods. A face is scaled and aligned to establish a baseline position prior to template extraction (Figure 4.1). Facial features are subsequently quantified to create a contour map of the position of individual facial features, and this is subsequently converted into a digital template (Ricanek & Boehnen, 2012).

During the matching process, pairs of digital templates are compared, and a numerical score is derived as a probabilistic measure of likeness (Garvie, Bedoya & Frankle., 2016, p. 9). The system developers determine the threshold of similarity for an identified match, and ideally, a match is confirmed by human decision-makers (Introna & Nissenbaum, 2010). There are a number of factors that are taken into consideration with respect to similarity thresholds, such as tolerance for false positives and negatives. Ultimately, these decisions should be context specific. It has also been

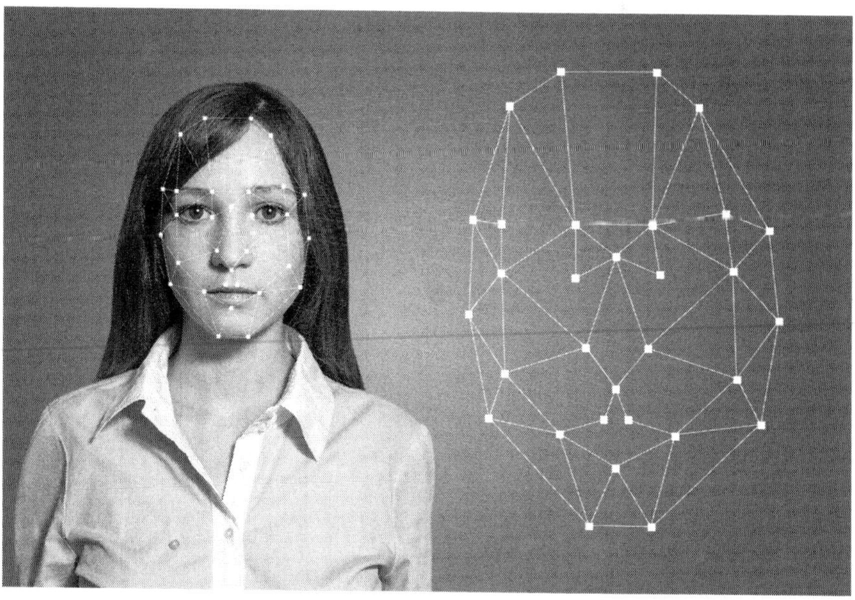

FIGURE 4.1 Facial contour mapping
Source: © artoleshko/Thinkstock.

argued that all matches should in the first instance be treated as potential false positives and verified by other information (Introna & Nissenbaum, 2010).

There are two main ways facial recognition can be used: verification and identification (Brey, 2004). Verification is undertaken through one-to-one matching, for example the comparison of faces with digital templates stored in identification documents or databases (Brey, 2004). Identification occurs through one-to-many searching, where databases are searched in a similar way to other biometrics for a facial template match. With these dual uses, facial recognition can be both 'targeted and public' in the case of confirming identification, yet it can also be 'generalised and invisible' in the case of widespread surveillance to identify and track unknown individuals (Garvie et al., 2016, p. 2). A subset of identification involves the use of a 'watch list', which is an additional way that one-to-many facial recognition can be used, for example, searching a crowd or public location for an individual or person on a watch list (Introna & Nissenbaum, 2010). As will be discussed later, this can occur through integration of facial recognition with already established CCTV systems and other emerging forms of video surveillance, such as body worn cameras.

Applications

There are several current and potential future applications of facial recognition in both the private and public sectors. Facial recognition is one of the least invasive forms of biometrics: it can be conducted from a distance, does not require the knowledge or consent of the subject, a physical sample is not required (such as with DNA identification) and it can be readily integrated with existing surveillance systems. It is also efficient: data analysis, searching and matching occur instantaneously. Huang and colleagues (2004) reviewed the major applications of facial recognition and identified the following categories: identification documents, access control, video surveillance, smart cards, law enforcement suspect alerts, facial image databases and human-computer interaction. This section reviews the major applications of facial recognition relevant to crime and security, and the emerging use of facial recognition in criminal investigations and trials.

Border control

Following the terrorist attacks on the United States in 2001, anyone entering the country was required to present a machine-readable biometric passport (Clark, 2011; Gates, 2006). Subsequently, the International Civil Aviation Organisation (ICAO) nominated facial recognition as the global standard for interoperable biometric passports (Huang, Xiong & Zhang, 2004; Clark, 2011). Facial recognition has subsequently been widely integrated with pre-existing methods of identification for border control purposes.

There has been a steady expansion of the use of face recognition technology for border control, expediting traveller processing around the world (Gold, 2014). Recently in the UK private sector, British Airways now uses facial recognition scanning at security screening, enabling travellers to board planes without showing identification documentation (Katz, 2017). More recently, there have been proposals for US Customs and Border Protection agents to use drones equipped with facial recognition technology to monitor the US border with Mexico, comparing images of individuals crossing the border with the Homeland Security Agency's Automated Biometric Identification System (IDENT) (Lee, 2017).

Facial recognition technology is now widely used around the world. In Australia, all international airports have SmartGate (or eGate) technology that automatically scans and compares travellers' faces with biometric identifiers stored within electronic passports (ePassports) (Department of Immigration and Border Protection, 2017). In late 2016, the Australian Government announced plans to phase out manual processing of passports by 2018. Instead of comparing a face with a facial template stored in an ePassport, the face would be compared with a facial template stored within the Australian Passport Office's database (Colley, 2016).

Video surveillance

A key capability of automated facial recognition is the capacity to identify a face in a large crowd. This could be a terrorist suspect in a sports stadium, a known shoplifter in a department store or a criminal walking through an airport. Perhaps the most significant application of facial recognition is the potential for integration into Smart CCTV systems. It enables real-time surveillance, identification and tracking of individuals through public places (Gates, 2011).

Smart CCTV was first used in 1998 by the London Metropolitan Police (Brey, 2004). Businesses in the United Kingdom can share CCTV footage directly with the police. Facial recognition technology can enable store owners to be notified when a shoplifter enters their store (BBC, 2015). In 2016, police used Smart CCTV to scan the faces of over one million people attending the Notting Hill Carnival (Boyle, 2016). Even more recently, it was reported that police in Wales scanned the faces of attendees at the 2017 Champions League football final in Cardiff, comparing them with a database of 500,000 persons of interest (Owen, 2017).

Facial recognition technology has also been used at large public events, including scanning the faces of attendees at the US Super Bowl (Chachere, 2001). At least five major police agencies within the United States, including those in New York and Los Angeles use real-time facial recognition from CCTV street cameras in public places. Real-time surveillance in public places also extends to cameras fitted at automatic teller machines (ATMs), in government vehicles, body worn cameras, drones and police robots. Facial recognition can potentially be integrated with any type of live video surveillance (Garvie et al., 2016).

Social media

Another notable application of facial recognition is the analysis of images taken from the Internet, particularly social media. Law enforcement can harvest images from Facebook, Twitter, LinkedIn, Google, among other websites, including dating sites. The significance of this capability is highlighted by the rapid expansion in the number of images uploaded to the Internet. For example, in 2012, Facebook alone held over 100 billion photos in its database, a figure estimated to increase by six billion photographs of facial images each month (Ricanek & Boehnen, 2012; Welinder, 2012).

Facebook actually uses facial recognition technology to tag photographs with users' names, and link images to personal information such as age, gender, location, contacts and political views (Bunn, 2014). Facebook users can also tag people in photographs, regardless of whether that person has a Facebook account themselves, and irrespective of whether or not they have consented to Facebook creating and storing their digital facial template (Bunn, 2014; de Andrade, Martin & Monteleone, 2013). Following these developments, the Hamburg Commissioner of Data Protection launched a legal challenge to Facebook's facial recognition tagging feature under German data protection and privacy laws (Mann & Smith, 2017). In 2012, the Irish Data Protection Commissioner audited Facebook's use of face recognition (Facebook's European headquarters are in Ireland), making a series of recommendations. In response, Facebook disabled the facial recognition tagging feature in Europe, deleted stored biometric information previously collected, and suspended the creation of new facial templates without prior active consent (Mann & Smith, 2017).

In 2016, an application known as FindFace was launched in Russia, allowing people to take photographs of anyone and search social media sites to identify them. The developers of this application claim that they have access to a database of over a billion photographs (Walker, 2016). The development also demonstrates how facial recognition technology can be used as a conduit to other large datasets that currently exist or are being compiled by private companies and public agencies. A person's face can now provide a link to a significant amount of data about an individual, including information relating to their digital and physical existence.

Facial expression analysis

A further extension of facial recognition technology is the analysis of facial expressions to infer internal emotional states, including deception (Tian, Kanade & Cohn, 2004; Gates, 2011). Facial expression analysis is defined as a 'computer system that attempt[s] to automatically analyse and recognise facial motions and facial feature changes from visual information' (Tian et al., 2004, p. 247). Facial expression analysis technology is not yet as advanced as automated facial recognition, however, it is an emerging area with many possible applications in crime and security. One of these is a modern alternative to polygraph lie detection. This technology can also be readily integrated with other forms of surveillance, access control and biometric technologies. For example, a partnership between the University of Arizona and the US Customs and Border Protection has developed an Automated Virtual Agent for Truth Assessments in Real-time (Border AVATARs) (National Center for Border

Security and Immigration, 2012). This combines both facial recognition technology and facial expression analysis. Further, the US Department of Homeland Security (2016) is developing Future Attribute Screening Technology (FAST) where a robotic interviewer asks a series of questions while assessing biometric information, including facial expressions and voice intonation, to detect deception.

Broader applications of facial expression analysis exist in advertising and marketing. This technology has been used in consumer research to establish which advertisements elicit the most positive emotional response (see. for example. Richards, 2016). Further, automated facial expression analysis was applied in a 2016 presidential debate in the United States to assess candidates' reactions to certain questions (Manning, 2016). In the future, automated facial expression analysis may become increasingly relevant for human and computer interaction. It could also be used to detect individuals with certain emotional states (for example aggravated or alert) in public places or transport systems, assisting the screening process.

Legal system developments

Currently, there have only been two reported cases of facial recognition being used in criminal investigations and presented as evidence at trial; both of these cases occurred in the United States. However, it is very likely that facial recognition has been used for intelligence purposes in the course of investigations to identify suspects or witnesses, and has not been made public either via the media or court documentation. Another possibility is that the technology has difficulty identifying unknown individuals, either as a consequence of the technology or because suspect images are not stored in databases enabling a match (Stroud, 2014).

The first legal case involving facial recognition evidence involved Charles Heard, who was accused of a murder committed during an armed robbery. The case went to trial in San Francisco in 2010. At trial, the defence sought to have facial recognition results admitted as evidence in an attempt to exculpate the accused, which was allowed by the judge (Jamison, 2010). Surveillance footage of the likely shooter was admitted as evidence, along with testimony from an expert who argued that the images of the shooter were not Charles Heard (Jamison, 2010). Although the jury was unable to agree on the identity of the shooter, Heard was convicted of first-degree murder as it was established that he had participated in the attempted robbery, resulting in the death, and was later sentenced to 25 years imprisonment (Nusca, 2011). In spite of the fact the extent to which the facial recognition evidence influenced the jury's decision-making in this case is unknown, a precedent was set: it was the first time facial recognition evidence was deemed admissible at trial. Questions remain regarding the extent to which facial recognition technology has gained acceptance within the scientific community, to the standard required in criminal trials. This issue will be considered further in Chapter 5.

The second case where facial recognition has been used at trial is a case involving a defendant named Pierre Martin. In 2014, Martin was convicted of two armed robberies on the Chicago train system committed the previous year. Using a facial recognition system, police searched a database of 4.5 million images and identified a match with Martin's previous arrest photograph (Nichols, 2014). Although the use of facial recognition was not mentioned at trial, due to the fact that Martin was also identified in a line up and admitted his guilt, it demonstrates the potential for this technology to be used in investigations to identify unknown suspects, where that individual's image is contained on an existing photographic database (Stroud, 2014).

As will be discussed further in Chapter 5, although facial recognition has the potential to form an important part of law enforcement investigations, automated facial recognition evidence has not as yet been admitted in a criminal trial, and issues with the admissibility of non-automated facial mapping evidence indicates it may not be straightforward. However, legal systems around the world are continually adapting to new technological developments in human identification.

Databases

There has been an exponential expansion in facial recognition databases around the world. In some cases, police information systems are used as a foundation for facial recognition databases. However, government and law enforcement agencies also routinely access, integrate and search existing databases, such as driver's licence and passport photograph databases. The prior existence of these high-quality digital images (that have been collected for the purpose of routine issuance of identification documentation) has enabled police to form large networks of biometric information that enables the searching of facial images and templates. Unlike DNA or fingerprint databases, these types of facial recognition databases formed from identification documents are comprised of individuals who have not previously been involved in the criminal justice system. Cases such as *Marper* in the European Court of Human Rights (ECtHR) have raised questions about the legality of retaining biometrics in this context, which may account for the limited presentation of this type of identification evidence at trial to date (Mann & Smith, 2017).

United States

The United States has a number of facial recognition databases at federal, state and local levels. The Next Generation Identification (NGI) system is the Federal Bureau of Investigation's (FBI's) primary multi-modal biometric database; it contains 100 million individual records of fingerprints, facial templates and photographs, iris scans and palm prints (Babcock, 2015). The NGI was integrated with, and replaced, the original Integrated Automated Fingerprint Identification System (IAFIS) (FBI, 2017a, 2017b) (discussed initially in Chapter 2 on fingerprinting). The NGI is the largest police information system in the world. It includes approximately 30 million photographs from 16.9 million individuals (US GOA, 2016). A sub-system known as the Interstate Photo System (IPS) comprises all photographs received by the FBI, along with accompanying ten-point fingerprints (FBI, 2017c). In addition to the IPS, the FBI's facial recognition database searches 16 state driver's licence databases (Garvie et al., 2016). The latest data available from the FBI indicates that 6,697 requests for facial recognition searches were made in February 2017.

In addition to the NGI-IPS, the US Department of Homeland Security operates the US Visitor and Immigrant Status Indicator Technology (US-VISIT) programme. The US-VISIT programme involves the collection of biometric information (ten fingerprints and digital photographs) from all non-US citizens entering the United States. Through this programme, the US Department of Homeland Security has developed a biometric repository of non-US citizens, known as the Automated Biometric Identification System (IDENT). US agencies that have access to IDENT include Customs and Border Protection, Immigration and Customs Enforcement, Coast Guard, Citizenship and Immigration Services, Department of State, Department of Defense, Department of Justice, law enforcement and the intelligence community. Access is also provided to US international partners including the International Criminal Police Organisation (INTERPOL) and Five Eyes partners (Australia, Canada, New Zealand and the United Kingdom) (US Department of Homeland Security, 2015).

It is important to emphasise that federal agencies such as the FBI and the Department of Homeland Security are not the only US agencies to maintain facial recognition databases. State and local police departments have also developed facial recognition databases and systems, which are equally as advanced as the NGI-IPS (Garvie et al., 2016). Approximately 30 states allow police and law enforcement to conduct facial recognition searches using driver's licence databases (Garvie et al., 2016), and it has been estimated that facial recognition searches apply to more than 117 million American adults, a number that is continually expanding. This equates to approximately half of all American adults having their photo identification in a facial recognition database (Garvie et al., 2016).

United Kingdom

The Police National Computer (PNC) contains photographs, DNA profiles, fingerprints and information about individuals who have been convicted of a criminal offence, recently arrested or are currently involved in legal proceedings. The PNC also contains information about firearms registration, and all vehicle registration details (Home Office, 2014). All police and intelligence agencies, including the Security Service and the Secret Intelligence Service, maintain access to the PNC. In response to a 2014 freedom of information request, the Home Office confirmed that there were 11,547,847 records (known as nominal records) on the PNC, approximately one million of which did not contain a criminal record, but were likely to include individuals currently facing prosecution, or that had previously been arrested but not charged (Home Office, 2015). It has recently been reported that police in England and Wales have created a facial recognition database containing up to 18 million photographs, including those who have never been charged with or convicted of a criminal offence, equating to approximately one third of the population (Hopkins & Morris, 2015).

Australia

The collection and implementation of facial recognition and other forms of automated biometric identification has also been expanding in Australia. This includes the introduction of biometric driver's licences by state governments, and at the federal level, the collection of biometric photographs through passport applications and immigration processing. This has enabled the introduction of biometric passports and automated facial recognition immigration clearance gates (Mann & Smith, 2017).

A significant development in Australia, and also internationally, is the introduction of a national facial recognition database – the National Facial Biometric Matching Capability (NFBMC) (Mann & Smith, 2017). The first stages of this system became operational in late-2016, with further expansion planned. The Capability enables a range of federal agencies to share and search facial templates. Participating agencies include the Department of Foreign Affairs and Trade (DFAT) (passport images), the Department of Immigration and Border Protection (DIBP) (visa images), the Australian Federal Police (AFP) and the Australian Security Intelligence Organisation (ASIO) (Attorney-General's Department, 2015a, 2015b). It is estimated that approximately half of the Australian population hold biometric passports, meaning that the NFBMC includes biometric facial templates of approximately 12 million citizens (DFAT, 2011). It is expected that other government agencies will also be able to conduct facial recognition searches in due course. For example, the Commonwealth Digital Transformation Agency (DTA) is currently designing and implementing a single digital identifier for access to all government systems and services, known as the Trusted Digital Identification Framework.

The newly formed Australian Criminal Intelligence Commission (ACIC, formerly the Australian Crime Commission and the CrimTrac agencies) is developing a multi-modal biometric database, known as the Biometric Identification Services (BIS) (ACIC, 2017). The BIS will include a national finger, palm and foot print database in addition to a national capability for facial recognition. It will also enable the fusion of multiple modes of biometric information and enable expansion with future developments in biometric modalities (ACIC, 2017). This will be discussed further in Chapter 5, in the context of other potential future developments in biometrics.

Emerging issues

Facial recognition is a less established form of biometric identification. It also has a number of characteristics that differentiate it from other forms of biometric identification techniques considered in this text. These include the potential for facial recognition databases to be integrated with other surveillance technologies, including CCTV; the public and the visible nature of faces enabling covert tracking; and the potential for widespread implementation. There are a range of emerging issues that are relevant to facial recognition. These relate to the accuracy of facial recognition systems, the potential for impacts on privacy and other civil liberties and the adequacy of oversight and accountability mechanisms. These issues are examined in this final part of this chapter.

Accuracy

Facial mapping, the precursor to biometric facial recognition, has been described as 'fraught with dangers' by legal experts, and lacking widely accepted procedures and reliability measures (Edmond et al., 2009, p. 337). An important issue relating to the use of facial mapping evidence, particularly in the context of criminal investigations and trials, is the fact that the distribution and frequency of facial features in the general population is unknown (Edmond et al., 2009). This means it is impossible to attest to the matches to any degree of certainty, or the probability that another individual has the same facial features. This situation can be contrasted with DNA evidence, where statistical estimates based on population datasets are routinely made in criminal trials, and are a foundation of the acceptance of the technique.

These concerns extend to the use of automated facial recognition technology, which has questionable accuracy, particularly when applied to non-cooperative and non-stationary subjects in uncontrolled conditions (see, for example, Grother, Quinn & Ngan, 2017). The risk of inaccuracy is increased when real-time CCTV footage or large databases are used (Garvie et al., 2016). Grother and colleagues (2017) studied facial recognition in video surveillance with non-cooperative subjects, finding that accurate face recognition among non-cooperative subjects in video surveillance was much more difficult than with portrait-style photographs. This is because non-stationary faces in video surveillance may appear across a range of resolutions (that may be impacted by magnification, field or depth of view), orientation or pose and lighting conditions. As the subjects of surveillance move their faces must be tracked through time, causing the blurring of images and rendering recognition more difficult. The impact of compressing images for storage may also negatively impact on recognition accuracy. Further, and perhaps most significantly, an individual can only be positively identified if their image has been enrolled into the face recognition database that an image is being compared with.

There are a number of factors that should be taken into consideration with regard to face recognition performance. It has been noted that although face recognition has demonstrated efficacy in small populations in controlled laboratory environments, it performs poorly in uncontrolled environments when individuals do not self-identity (or perhaps do not wish to be identified) (see, for example, Introna & Nissenbaum, 2010). Performance is also dependent on the environment, age of images for comparison, similarity of cameras used for image enrolment and comparison and the size of any associated databases. It has also been noted that there is currently limited publicly available data on operational identification (Introna & Nissenbaum, 2010). There appears to be a need for further research into the accuracy of facial recognition systems and its use under various environmental conditions.

Changes in an individual's face over time may also lead to false positive or negative matches. This could include changes as a consequence of aging, cosmetic surgery, make up or weight gain. Facial recognition conducted via video surveillance requires that faces are visible and not concealed by hair, glasses, headscarves or other head wear. There have also been attempts to develop anti-surveillance clothing to spoof or defeat facial recognition technology in video surveillance (Samuels, 2017).

Even human match confirmation – either with traditional facial mapping techniques or examination of similarity of photographs – has been found to be inaccurate approximately half of the time. There is also limited capacity within law enforcement agencies to conduct manual confirmation of matches (Garvie et al., 2016). Inaccurate identification has a range of consequences, including the potential to involve innocent people in law enforcement investigations and subject them to unwarranted attention.

To further compound concerns associated with the accuracy of facial recognition systems, the identification of ethnic minorities are subject to greater risk of inaccuracy. This is due to the fact that algorithms that compare facial templates may skew or influence the types of faces that are identified. As some facial recognition systems in the United States draw from police databases containing mug shot images of a disproportionate number of African-American individuals, it has been argued that facial recognition technology will mostly affect certain racial groups, which are already subject to disproportionately high rates of police attention, arrest and incarceration (Garvie et al., 2016).

Given these issues and concerns about facial recognition, it is interesting that police around the world have quickly moved to implement large facial recognition systems, and in some cases, concern has been expressed by oversight bodies. The US Government Accountability Office (GAO) raised concerns about the FBI's limited testing of their facial recognition system for accuracy (Garvie et al., 2016).

Privacy

There are a number of potential privacy issues associated with facial recognition technology that extend beyond other forms of biometric identification. Faces are difficult to hide and alter, and are linked to an individual's physical existence, meaning that individuals can be tracked through public space (Buckley & Hunter, 2011). This is especially relevant in the context of integration with existing and new pubic surveillance systems such as Smart CCTV, drones, vehicle cameras and body worn cameras. There are also a number of questions associated with the way in which facial templates are appropriated from other pre-existing administrative or identification databases, irrespective of whether individuals have a criminal conviction, and the necessity and proportionality of their collection, retention and use (Mann & Smith, 2017).

There is the potential for facial recognition to be used to surveille public protests and have a chilling effect on free speech and other forms of civil rights participation (Garvie et al., 2016). Limited legal or administrative protections have been introduced to prevent the use of facial recognition identification and surveillance at public protests. For example, of 52 agencies who reported using facial recognition in the United States, only 1 has prohibited the use of facial recognition for the purposes of identifying individuals engaging in free speech and civil protest (Garvie et al., 2016). Access to driver's licence and passport databases may also be a concern if it is occurring without the requirement for a warrant or court order, or where there are no requirements for police to have reasonable suspicion prior to conducting a facial recognition search (Garvie et al., 2016). These potential issues must be balanced against the significant public interest in investigating and prosecuting crime. Further, it is common for regulatory gaps to exist when technology is first introduced, and it would be expected that an appropriate balance will be struck to manage these issues as facial recognition becomes even more established.

Regulation

Facial recognition databases should only be created in environments where sufficient oversight, and regulatory measures have been established (Lochnew, 2013). Some government agencies are not transparent about their use of facial recognition, or associated practices. It has also been found that major facial recognition systems in the United States are not audited for misuse, with some not subject to auditing at all (Garvie et al., 2016). Only 9 of the 52 agencies surveyed in the United States participate in some form of record keeping and auditing concerning their use of facial recognition systems. The Government Accountability Offices (GAOs) investigated the FBI's compliance with US privacy protections, finding that the agency did not release privacy impact assessments (PIAs) or publish other relevant documentation required with facial recognition systems until after their review was completed, and also raised concerns about the limited accuracy of the testing that was being conducted (Garvie et al., 2016).

As has been discussed, in the United Kingdom a Commissioner for the Retention and Use of Biometric Material was established in 2012 with a mandate to ensure responsible governance of the retention and use of biometric information. However, under current legislation, the Biometrics Commissioner's powers do not extend to facial recognition: only to DNA and fingerprints. A recent report on current and future uses of biometrics in the United Kingdom recommended that the responsibilities of the Biometrics Commissioner be extended to 'cover, at a minimum, the police use and retention of facial images' (House of Commons Science and Technology Committee, 2015, p. 34). The Biometrics Commissioner has also expressed concern about insufficient oversight of facial recognition technology, and the lack of regard for broader ECtHR decisions concerning the retention of biometric information of individuals who have not been convicted of a criminal offence (MacGregor, 2016).

It may be appropriate to require that legislation be enacted ensuring that police must have a reasonable suspicion of criminal conduct before searching facial recognition databases, and only when relevant to the investigation of serious crimes. More stringent requirements have been proposed by Garvie and colleagues (2016). They argue that facial recognition databases should comprise police arrestee photographs rather than driver's license databases, and that databases should be regularly audited to ensure that individuals who do not have a criminal record are not included. They argue for increased transparency around policies governing the use of facial recognition, public reporting on the number of images held in databases and number of searches conducted and accuracy testing. As was acknowledged earlier, it is important that a balance is struck between these matters, whilst ensuring that police have the resources necessary to deter, investigate and prosecute criminal activity.

References

Adler, A. & Schuckers, M. (2007). Comparing human and automatic face recognition performance, *IEEE Transactions on Systems, Man, and Cybernetics – Part B: Cybernetics, 37*, 1248.

Attorney-General's Department. (2015a). Preliminary privacy impact assessment of the National Facial Biometric Matching Capability – Interoperability Hub. Retrieved from https://www.ag.gov.au/rightsandprotections/identitysecurity/pages/biometrics.aspx

Attorney-General's Department. (2015b). *Biometrics*. Retrieved from https://www.ag.gov.au/RightsAndProtections/IdentitySecurity/Pages/Biometrics.aspx

Australian Criminal Intelligence Commission (ACIC). (2017). Biometric identification service. Retrieved from https://www.acic.gov.au/our-services/biometric-matching/biometric-identification-services

Australian Privacy Foundation. (2016). *Biometrics*. Retrieved from https://www.privacy.org.au/Papers/Biometrics-0804.html

Amoore, L. (2006). Biometric borders: Governing mobilities in the war on terror. *Political Geography 25*, 336.

Babcock, E.J. (2015). *Privacy Impact Assessment for the Next Generation Identification (NGI) Interstate Photo System*. Retrieved from https://www.fbi.gov/foia/privacy-impact-assessments/interstate-photo-system

BBC. (2015, 16 December). Facewatch 'thief recognition' CCTV on trial in UK stores. *BBC News*. Retrieved from http://www.bbc.com/news/technology-35111363

Boyle, D. (2016, 27 August). Police to scan 1 million people with new automatic facial recognition software in bid to beat crime at Notting Hill Carnival. *The Daily Mail*. Retrieved from http://www.dailymail.co.uk/news/article-3761236/Police-scan-1million-people-new-automatic-facial-recognition-software-bid-beat-crime-Notting-Hill-Carnival.html

Brandom, R. (2017, 18 April). Facial recognition is coming to US airports fast-tracked by Trump. *The Verge*. Retrieved from http://www.theverge.com/2017/4/18/15332742/us-border-biometric-exit-facial-recognition-scanning-homeland-security

Brey, P. (2004). Ethical aspects of facial recognition systems in public places. *Journal of Information, Communication and Ethics in Society 2*, 97.

Bunn, A. (2014). Facebook and face recognition: Kinda cool, kinda creepy. *Bond Law Review 25*(1), 35.

Buckley, B. & Hunter, M. (2011). Say cheese! Privacy and facial recognition. *Computer Law and Security Review 27*, 637–640.

Chachere, V. (2001, 13 February). Biometrics used to detect criminals at super bowl. *ABC News*. Retrieved from http://abcnews.go.com/Technology/story?id=98871

Clarke, S.R. (2011). Balancing privacy and security in the Australian passport system. *Deakin Law Review 16*(2), 325.

Colley, A. (2016, 21 October). Govt wants to remove passports from border processing. *IT News*. Retrieved from https://www.itnews.com.au/news/govt-wants-to-remove-passports-from-border-processing-439785

de Andrade, N.N.G., Martin, A., & Monteleone, S. (2013). 'All the better to see you with my dear': Facial recognition and privacy in online social networks. *IEEE Computer and Reliability Societies, May/June*, 21–28.

Department of Foreign Affairs and Trade (DFAT). (2011). *Annual Report 2010–2011*. Retrieved from http://dfat.gov.au/about-us/publications/corporate/annual-reports/annual-report-2010-2011/index.html

Department of Immigration and Border Protection. (2017). What is the face recognition technology used in arrivals SmartGate? *Australian Government Factsheet*. Retrieved from https://www.border.gov.au/FAQs/Pages/What-is-the-face-recognition-technology-used-in-arrivals-SmartGate.aspx

Edmond, G., Biber, K., Kemp, R. & Porter, G. (2009). Law's looking glass: Expert identification evidence derived from photographic and video images. *Current Issues in Criminal Justice 20*(3), 337.

Federal Bureau of Investigation (FBI). (2017a). *Next Generation Identification (NGI)*. Retrieved from https://www.fbi.gov/services/cjis/fingerprints-and-other-biometrics/ngi

Federal Bureau of Investigation (FBI). (2017b). *NGI Monthly Fact Sheet*. Retrieved from https://www.fbi.gov/file-repository/ngi-monthly-fact-sheet/view

Federal Bureau of Investigation (FBI). (2017c). *Next Generation Identification: Implementing the Future of Identification and Investigative Services Flyer*. Retrieved from https://www.fbi.gov/file-repository/next-generation-identification-ngi-flyer.pdf/view

Franceschi-Bicchierai, L. (2015, 19 March). US Customs quietly launches facial recognition experiment at DC airport. *Motherboard*. Retrieved from https://motherboard.vice.com/en_us/article/us-customs-quietly-launches-facial-recognition-experiment-at-dc-airport

Garvie, C., Bedoya, A.M. & Frankle, J. (2016). The perpetual line-up: Unregulated police face recognition in America. *Georgetown Law Centre on Privacy and Technology Report*. Retrieved from https://www.perpetuallineup.org

Gates, K.A. (2002). Wanted dead or digitised: Facial recognition technology and privacy. *Television and New Media 3*(2), 235.

Gates, K.A. (2006). Identifying the 9/11 'faces of terror'. *Cultural Studies 20*, 4.

Gates, K.A. (2011). *Our Biometric Future: Facial Recognition Technology and the Culture of Surveillance*. New York: New York University Press.

Gold, S. (2014). Biometrics at the border. *Biometric Technology Today, October*, 5.

Grother, P., Quinn, G. & Ngan, M. (2017). Face in video evaluation (FIVE): Face recognition of non-cooperative subjects. *National Institute of Standards and Technology, US Department of Commerce*. Retrieved from nvlpubs.nist.gov/nistpubs/ir/2017/NIST.IR.8173.pdf

Home Office, UKGovernment. (2014). *Police National Computer*. Retrieved from https://www.gov.uk/government/publications/police-national-computer-pnc

Home Office, UKGovernment. (2015). FOI release: Nominal criminal records on the police national computer. Retrieved from https://www.gov.uk/government/publications/

nominal-criminal-records-on-the-police-national-computer/nominal-criminal-records-on-the-police-national-computer

Hopkins, N. & Morris, J. (2015, 3 February). Innocent people on police photos database. *BBC News*. Retrieved from http://www.bbc.com/news/uk-31105678

House of Commons Science and Technology Committee. (2015). *Current and Future Uses of Biometric Data and Technologies*. London: UK Parliament. Retrieved from http://www.parliament.uk/business/committees/committees-a-z/commons-select/science-and-technology-committee/inquiries/parliament-2010/current-and-future-uses-of-biometric-data-and-technologies

Huang, T., Xiong, Z. & Zhang, Z. (2004). Face recognition applications. In Li, S. & Jain, J. (eds), *Handbook of Face Recognition* (pp. 371–390). London: Springer-Verlag.

Introna, L. & Nissenbaum, H. (2010). Facial recognition technology: A survey of policy and implementation issues. *The Department of Organisation, Work and Technology, Lancaster University Working Paper*. Retrieved from http://www.research.lancs.ac.uk/portal/en/publications/facial-recognition-technology-a-survey-of-policy-and-implementation-issues(43367675-c8b9-4644-90f2-86815cc8ea15).html

Irish Data Protection Commissioner. (2012). Report of re-audit: Facebook Ireland Ltd. Retrieved from https://www.dataprotection.ie/docs/21-09-12-Facebook-Ireland-Audit-Review-Report/1232.htm

Jain, A.K., Ross, A., & Prabhakar, S. (2004). An introduction to biometric recognition. *IEEE Transactions on Circuits and Systems for Video Technology, Special Issue on Images and Video-Based Biometrics 14*(1), 1.

Jamison, P. (2010, 14 July). Facial profiling: Will face-recognition technology get an accused killer off the hook? *SF Weekly*. Retrieved from http://archives.sfweekly.com/sanfrancisco/facial-profiling/Content?oid=2177840&storyPage=5

Katz, B. (2017, 24 March). British Airways starts scanning faces to enable faster boarding. *Bloomberg Technology*. Retrieved from https://www.bloomberg.com/news/articles/2017-03-24/british-airways-starts-scanning-faces-to-enable-faster-boarding

Keenan, M. (2015). New $18.5 million biometrics tool to put a face to crime. Retrieved from https://www.ministerjustice.gov.au/Mediareleases/Pages/2015/ThirdQuarter/9-September-2015-New-$18–15-million-biometrics-tool-to-put-a-face-to-crime.aspx

Lee, J. (2017, 12 April). CBP seeking face recognition drones to monitor border. *Biometric Update*. Retrieved from http://www.biometricupdate.com/201704/cbp-seeking-face-recognition-drones-to-monitor-border

Lochnew, S.A. (2013). Saving face: Regulating law enforcement's use of mobile facial recognition technology and iris scans. *Arizona Law Review 55*, 201.

MacGregor, A.R. (2016). *Annual Report 2015: Commissioner for the Retention and Use of Biometric Material*. Retrieved from https://www.gov.uk/government/uploads/system/uploads/attachment_data/file/507104/54496_Biometrics_Commissioners_Report_Print_Ready__3_.pdf

Mann, M. & Smith, M. (2017). Automated facial recognition technology: Recent developments and approaches to oversight. *University of New South Wales Law Journal 40*(1), 121.

Manning, A. (2016, 27 September). Presidential debate: Here's how the candidates really reacted. *Vocativ*. Retrieved from http://www.vocativ.com/362917/presidential-debate-clinton-trump-candidates-reaction-analysis

National Center for Border Security and Immigration. (2012). *Annual Report – Year 4: July 2011 to June 2012*. Retrieved from http://www.borders.arizona.edu/cms/sites/default/files/BORDERS-YR4-AnnualReport-Revised-121220-FINAL.pdf

NEC. (2015). NEC facial recognition helps NT Police solve cold cases and increase public safety in Australia. Retrieved from http://au.nec.com/en_AU/press/201509/nec-facial-recognition-increases-public-safety-in-australia.html

Neilsen, M.A. (2015). *Migration Amendment (Strengthening Biometrics Integrity) Bill 2015 Bills Digest*. Canberra: Australian Parliament.

Nichols, S. (2014, 10 June). Facial recognition tech convicts man in Chicago robbery case. *The Register*. Retrieved from https://www.theregister.co.uk/2014/06/10/facial_recognition_gets_its_first_conviction_in_chicago_robbery_case

Nusca, A. (2011, 1 February). Biometrics valid evidence in trial, judge rules: A San Francisco judge ruled that biometric facial recognition could be submitted as legal evidence in a trial. *ZDNet*. Retrieved from http://www.zdnet.com/article/biometrics-valid-evidence-in-trial-judge-rules

Owen, G. (2017, 26 April). British cops will scan every fan's face at the Champions League final. *Motherboard*. Retrieved from https://motherboard.vice.com/en_us/article/british-cops-will-scan-every-fans-face-at-the-champions-league-final

Pampus, J. & Weber, F. (1998). Facial recognition: An overview. *Information Security Technical Report 3*(1), 40.

Ricanek, K. & Boehnen, C. (2012). Facial analytics: From big data to law enforcement. *IEEE Computer Society, September*, 95.

Ricanek, K. (2014). Beyond recognition: The promise of biometric analytics. *IEEE Computer Society, September*, 87.

Robertson, J. (2017, 8 March). Fears over trial of '1984' surveillance system that anticipates antisocial acts. *The Guardian*. Retrieved from https://www.theguardian.com/world/2017/mar/08/queensland-council-trials-iomniscient-surveillance-to-anticipate-antisocial-acts

Richards, K. (2016, 12 February). Facial-tracking technology shows these 5 Super Bowl ads were the most engaging. *Ad Week*. Retrieved from http://www.adweek.com/brand-marketing/facial-tracking-technology-says-these-5-super-bowl-50-ads-were-most-engaging-169636

Samuels, G. (2017, 5 January). Anti-surveillance clothing unveiled to combat facial recognition technology. *The Independent*. Retrieved from http://www.independent.co.uk/news/science/anti-surveillance-clothing-facial-recognition-technology-hyperface-adam-harvey-berlin-facebook-apple-a7511631.html

Stroud, M. (2014, 8 August). Did Chicago's facial recognition system catch its first crook? *The Verge*. Retrieved from http://www.theverge.com/2014/8/8/5982727/face-wreck-how-advanced-tech-comes-up-short-for-police

Tian, Y., Kanade, T., & Cohn, F. (2004). Facial expression analysis. In Li, S. & Jain, J. (eds), *Handbook of Face Recognition* (pp. 247–275). New York: Springer-Verlag.

UKGovernment. (2017). Automated facial recognition solution. Retrieved from https://www.contractsfinder.service.gov.uk/Notice/0953e962-4d12-46bf-b298-5b88623c8e05?p=@NT08=UFQxUlRRPT0=NjJ

US Department of Homeland Security. (2015). *DHS/NPPD/Privacy Impact Assessment-002 Automated Biometric Identification System (IDENT)*. Retrieved from https://www.dhs.gov/publication/dhsnppdpia-002-automated-biometric-identification-system-ident

US Department of Homeland Security. (2016). *Future Attribute Screening Technology Fact Sheet*. Retrieved from https://www.dhs.gov/publication/future-attribute-screening-technology

US Government Accountability Office (US GOA). (2016). *Face Recognition Technology: FBI Should Better Ensure Privacy and Accuracy*. Report to the Ranking Member, Subcommittee on Privacy, Technology and the Law, Committee on the Judiciary. Retrieved from https://www.gao.gov/products/GAO-16-267

Valentine, T. & Davis, J. (2016). *Forensic Facial Identification: Theory and Practice of Identification from Eyewitnesses ad CCTV*. Chichester: Wiley-Blackwell.

VisionBox. (2016, 15 March). New York's John F. Kennedy International Airport deploys VisionBox biometric passport authentication technology. Retrieved from http://www.

vision-box.com/news/new-yorks-john-f-kennedy-international-airport-deploys-vision-box-biometric-passport-authentication-technology

Walker, S. (2016, 17 May) Face recognition app taking Russia by storm may bring end to public anonymity. *The Guardian*. Retrieved from https://www.theguardian.com/technology/2016/may/17/findface-face-recognition-app-end-public-anonymity-vkontakte

Welinder, Y. (2012). Face recognition privacy in social networks under German law. *Communications Law Bulletin 31*(1), 5.

Wilson, D. (2007). Australian biometrics and global surveillance. *International Criminal Justice Review 17*(3), 207.

5

NEW AND DEVELOPING FORMS OF BIOMETRIC IDENTIFICATION

Introduction

Scientific advancement and ongoing security requirements have led to the development of new forms of biometric identification. Biometric identification technologies are improving and becoming less expensive, allowing for wider adoption and increased accuracy. This chapter describes the introduction of new and emerging biometric modalities, including advancements in physiological (first generation) and behavioural (second generation) biometric modalities. First, new developments in physiological forms of identification are considered, including ear, vascular, ocular (retina and iris) and voice recognition. Subsequently, the developing field of behavioural biometrics is reviewed, including a discussion of gait recognition, keystroke dynamics and cognitive biometrics. The principles, application and issues associated with each new biometric modality are outlined, demonstrating a range of possible applications in crime and security, including advantages and disadvantages that should be considered. Concerns associated with the security of new biometric systems are examined, along with other related issues.

First generation biometrics

The first generation of biometrics that have been discussed so far in this book are derived from purely physiological traits: fingerprints, DNA, facial structure. Physiological traits used for the purposes of biometric identification are known as 'first generation' biometrics. The first generation biometrics examined throughout this section include those based on the ears, blood vessels, eyes and voice. Second generation biometrics, also known as behavioural biometrics, measure learned behaviours. The second generation biometrics considered in this chapter include gait, keystroke and cognitive biometrics. Voice recognition is a unique biometric identifier that combines both physiological and behavioural characteristics. In contrast with first generation biometrics, second generation biometrics are easier to change and mimic, presenting additional issues relating to accuracy, security and reliability.

Ear recognition

Principles

Ear recognition involves the automated extraction and comparison of the anatomical features of the human ear for the purposes of identification and verification (Pun & Moon, 2004). The use of the human ear to identify individuals was first suggested by French criminologist Alphonse Bertillon (1853–1914), who used measurements of the ear in his Bertillonage system to identify recidivists, and the first system of ear recognition was developed in 1949, integrating 12 measurements of the outer ear (Abaza et al., 2013).

Ear recognition involves the extraction and comparison of the unique features of the outer ears. Human ears can be used as unique identifiers because human ear growth is proportional to age, does not change radically across the lifespan and is not influenced by changes in expression (Anwar, Ghany &, 2015). Human ears are unique among individuals, including identical twins, making them a suitable biometric (Pflug & Busch, 2012). Researchers have obtained 98 per cent accuracy in identification using ear recognition in controlled environments (Anwar et al., 2015).

Application and issues

Ear recognition is a developing field and has not been as widely implemented as other emerging biometric modalities (Pun & Moon, 2004), and research into new methods of ear recognition, for example, three-dimensional imaging is ongoing (Ali & Islam, 2013). Ear recognition can be integrated with other forms of biometric identification such as facial recognition in order to address accuracy issues in a single biometric modality (Wang et al., 2012). As ear recognition can identify subjects at a distance, it can be implemented into smart closed circuit television (CCTV) surveillance systems or used for the purposes of border control (Pflug & Busch, 2012). It has been reported that the US Immigration and Naturalization Service (USINS) specifies the right ear should be visible in identification photographs. This may indicate that ear recognition is currently used for border control identification in the United States, although this is unclear from open source literature (Kumar & Srinivasan, 2014).

Ear recognition shares many of the issues associated with facial recognition, including the potential impact of lighting, head rotation and the potential for a subject to cover their features. As ears may easily be hidden by hair or headwear, ear recognition would be more suited to the identification of cooperative subjects. As ears have a smaller surface area, head rotation is likely to have greater impact on the accuracy of identification (Pun & Moon, 2004).

One of the main advantages of ear recognition over facial recognition is that ear recognition requires a smaller image size, and therefore of similar resolution meaning that it requires less memory for image storage and processing (Pun & Moon, 2004). Further, in comparison with faces, ears have greater uniformity in colour distribution, and less variability as a result of changes in expression (Pun & Moon, 2004). It is believed that ear recognition is the most promising biometric modality to be combined with facial recognition systems, as it can provide additional information on both sides of the face (Abaza et al., 2013). When combined with facial recognition systems, ear recognition can provide further contextual information to offset some of the adverse impacts and barriers to facial recognition accuracy such as illumination, pose and change of expression (Wang et al., 2012). In comparison with other modalities of biometric identification, ear recognition does not require specialist imaging equipment, is contactless, less invasive and stable over time (Pun & Moon, 2004).

Vascular pattern recognition

Principles

Vein or vascular pattern recognition (VPR) involves the imaging, extraction and comparison of subcutaneous vascular networks located under the skin, usually in hands and fingers, for the purposes of verifying identity. Vein recognition differs from other forms of first generation biometrics as it uses a *non-visible* physiological characteristic for the purposes of authentication. An infrared light source and infrared camera are used to identify the vein pattern concealed under the skin. The haemoglobin present in blood reflects the infrared light and provides visibility of the blood vessels (Smorawa & Kubanek, 2015). After the structure of the veins is obtained, vein recognition involves the same comparison methods as biometric fingerprinting (Smorawa & Kubanek, 2015). Like fingerprints, the pattern of blood vessels is unique and stable across an individual's life. The use of finger and palm veins consistently demonstrates very high rates of identification accuracy in comparison with fingerprint identification (Benziane & Benyettou, 2016).

Application and issues

The main application of vein recognition is identity verification, particularly in access control situations, such as border security and banking at automatic teller machines. Collinson (2014) reports that the use of vein patterns in fingers is a common method of identification for Japanese bank customers. Infrared vein recognition cameras can be mobile and integrated into other biometric systems. In the future, it is expected that vein recognition technology would be suitable for integration with computer, mobile telephone and vehicle entry applications (Sandhu & Kaur, 2015).

As vein networks are not externally visible, it is very difficult to change or copy (Sandhu & Kaur, 2015). Further, vein recognition is contactless and non-invasive, and given the high level of security it offers, and its convenience, it is expected that its applications will increase over time. Disadvantages of vein recognition include the necessity of infrared cameras and the potential for the ambient environmental temperature and medical conditions to affect its accuracy (Benziane & Benyettou, 2016).

Occular biometrics

Principles

Ocular biometrics involve the extraction and comparison of the anatomical features of the eye. The main structures of the eye include the cornea, lens, optic nerve, retina, pupil and iris. To date, iris recognition is the most common application; however, increasing attention is being paid to retina recognition, particularly as it is considered to be one of the most secure biometric modalities (Nigam, Vatsa & Singh, 2015).

Retina recognition utilises the vascular pattern of the retina. A retinal scanner is used to illuminate a region of the retina through the pupil, capturing the vascular pattern of the retina (Borgen, Bours & Wolthusen, 2009). Retina recognition is believed to be the most secure form of biometrics, due to its stability, uniqueness and the fact that it is very difficult to copy and replicate the vascular pattern of the retina (Waheed et al., 2016). However, retina recognition has not been widely adopted to date because of the cost of the highly specialised equipment and high levels of cooperation required of users. Due to the high level of security it offers, retina recognition has most commonly been implemented in military and nuclear facilities (Nigam et al., 2015).

The iris is the coloured part of the eye situated between the pupil and the sclera (the white part of the eye) (Ives et al., 2013). It consists of a series of layers of blood vessels that form distinct and complex patterns. The unique lattice of the iris forms at eight months gestation, and remains stable throughout an individual's life, with the exception of disease or trauma. The iris is therefore more stable than other forms of biometric modalities, such as faces and fingerprints. Iridial patterns are not only unique for each individual, but also between each eye (Pierscionek, Crawford & Scotney, 2008). Iris recognition involves the capture, extraction and comparison of these patterns. As is the case with other forms of biometric identification, the main stages of iris recognition include image acquisition, feature or pattern extraction, template generation and comparison (Ives et al., 2013).

Iris recognition is a non-invasive procedure: multiple frames of high-resolution grey scale images are required, illuminated with infrared or, in some cases, visible light (Borgen et al., 2009). Ongoing development of sensor technology has enabled more flexibile iris recognition systems; however, current iris recognition technology limits collection distances to approximately 30 centimetres (Nigam et al., 2015).

Application and issues

The adoption of biometric iris recognition is increasing with developments in technology and increased consumer demand both in public and commercial sectors. There has been a large-scale adoption of iris recognition technology for security applications, particularly in the areas of border control (Borgen et al., 2009). Iris scanners are currently deployed in many major airports around the world (Pierscionek et al., 2008). The United Arab Emirates (UAE) uses iris recognition at land, sea and air border points, and maintains a database of 1.1 million iris templates, one of the largest databases of its kind in the world. Between 2010 and 2013, the UAE conducted iris searches for 10.5 million individuals, identifying 124,435 people who were attempting to return to the UAE with forged identification documents (Ives et al., 2013). It has also been suggested that iris recognition could be used to reduce electoral fraud. There have been trials of iris based recognition for voter registration systems in African countries (Bowyer, Ortiz & Sgroi, 2015). Lecher & Brandom (2016) report that a Federal Bureau of Investigation (FBI) pilot programme, beginning in 2013, has collected iris scans from over 434,000 residents of the United States. This information is obtained via information-sharing arrangements with local law enforcement agencies, US Border Patrol and the US military.

Iris and retina recognition have the advantage of high levels of accuracy; however, there are some questions about the stability of iris patterns over time, due, for example, to medication, surgery, disease and aging (Pierscionek et al., 2008). These include glaucoma, macular degeneration, cataracts and pathological angiogenesis (Borgen et al., 2009). An area of potential future development in iris recognition technology is the capture of iris images while a subject is in motion or is uncooperative (Colores et al. 2011), and the development and use of mobile platforms to capture iris images (Barra et al., 2015). A recent development in this area are walk-through systems that can capture iris images without the subject stopping or removing glasses (Ives et al., 2013). Studies have even shown some success in capturing iris patterns while subjects were wearing sunglasses (Latman & Herb, 2013). It is anticipated that in the future there will be greater adoption of walk-through iris recognition in transportation, immigration and government facilities, as well as the development of mobile or portable iris recognition systems (Ives et al., 2013).

In mid-2017, it was reported that the iris recognition system on widely used smartphones could easily be circumvented by using the night mode of a digital camera to take an infrared picture of the phone user's eyes from a moderate distance, print-out a life size picture and hold it in front of the phone (Meyer, 2017). Examples such as this highlight the importance of further investment in measures to counter circumvention, otherwise the substantial amounts spent on research and development may be undermined upon release of the technology.

Voice recognition

Principles

Voice recognition applies the individual characteristics of the human voice as a biometric identifier through the extraction and comparison of voice samples

(Galka, Masior & Salasa, 2014). Unlike other forms of biometric analysis and identification, this biometric involves a combination of both physiological and behavioural characteristics. One of the main advantages of speaker recognition is that there are several voice characteristics that can be analysed and compared, enabling a high level of accuracy in identification (Morgen, 2012).

The physiological characteristics of human voices relate to anatomical differences in the biological structure of the vocal tract. There are three main areas of the vocal tract that are known as the infraglottic, glottal and supraglottic areas that influence voice production. When a person speaks, the effects of air pressure, muscle tension and elasticity of the vocal folds are modulated to create different sounds. The frequency of sound pressure patterns are analysed for biometric identification purposes. The behavioural features of voices are influenced by how an individual has learned to speak, including their vocabulary, accent, intonation, pitch, pronunciation and conversational patterns (Mazaira-Fernandez, Alvarez-Marquina & Gomez-Vilda, 2015).

Application and issues

As with the other physiological forms of biometrics discussed throughout the text, applications for voice recognition include the identification of unknown individuals, and verification of identity. Voice recognition has applications in banking and government service provision via telephone, or access control (Kaman et al., 2013). It also has applications in audio forensics, where other types of biometric identification are not available, such as a case where a suspect is wearing a mask. Voice recognition can also be used to identify individuals on social media videos or intercepted phone conversations (Mazaira-Fernandez et al., 2015).

There are a wide range of applications and technologies utilising voice recognition, such as personal computers, mobile devices and social robotics. For example, the Australian Taxation Office introduced voluntary 'voiceprint' technology for callers to verify their identities when contacting the agency. By 2016, more than 1.5 million Australians had been enrolled within the voiceprint programme (Nuance Communications, 2016).

The main issues associated with voice recognition relate to the variability in an individual's voice as a consequence of their mood, health or the aging process, which can all impact on various characteristics of the voice (Mazaira-Fernandez et al., 2015). There is also the potential for issues associated with ambient or environmental noise, and distortions (Chenafa et al., 2008). In response to concerns that individuals may be able to disguise their voices through electronic means, researchers are currently developing new methods to successfully identify electronically disguised voices (Wu, Wang & Huang, 2014). Key advantages of voice recognition include its non-invasive nature, the fact that it does not require specialised hardware, aside from a traditional microphone, and that it can be conducted remotely via a telephone or the Internet (Kaman et al., 2013).

Second generation biometrics

In comparison with first generation physiological biometrics, second generation biometrics concern the analysis of learned behaviour, and are described as *behavioural biometrics*. The development of behavioural biometrics is more recent than physiological biometrics, and has been described as second generation biometric identification. They have been developed from an analysis of learned behaviour, and are more likely to change over time than physiological identifiers. Notwithstanding issues in relation to stability, accuracy and reliability, behavioural biometrics have a wider range of applications in comparison with physiological biometrics. The biometrics that will be considered here include gait recognition, keystroke dynamics and cognitive biometrics.

Gait recognition

Principles

Gait recognition is situated within the broader field of human motion analysis, involving the examination and comparison of human kinesiology (Neves et al., 2016). Everyone has a unique and regular pattern of motion when walking, relating to the movement of their limbs. Gait recognition involves the measurement, analysis and comparison of human movement made by an individual when they walk (Chaurasia et al., 2015).

Gait recognition is one of the more recent forms of biometric identification to be developed and coincides with computer processing advancements (Nixon & Carter, 2006). There are a number of stages involved in gait recognition. These involve capturing a walking sequence captured from video input, creating a movement silhouette and the extracting of static and dynamic features across a sufficient period of time. Movement silhouettes are transformed into a gait cycle, depicting a sufficient walking period to be used for the purposes of comparison and identification (Indumathi & Pushparani, 2016). In addition to walking patterns, gait recognition systems can also collect and analyse the physical appearance of individuals such as the height, length of limbs, shape and size of torso (Zhang, Hu & Wang, 2011). Identification therefore occurs through both shape (physiological features) and motion (behavioural features) (Nixon & Carter, 2006; Choudhury & Tjahjadi, 2013).

Gait recognition technology has reached 90 per cent accuracy in identification, provided there are analogous environmental conditions in the comparison footage. However, the walking surface and clothing can influence the recognition rates (Nixon & Carter, 2006). Different camera viewpoints can also improve the rate of identification accuracy. For example, in a study of gait recognition published in 2016, Bouchrika and colleagues obtained an identification accuracy rate of 73 per cent for gait features extracted from individual camera viewpoints, which could be increased to an identification accuracy rate of 92 per cent with cross-camera matching. Other research groups have achieved a similar average identification accuracy rate utilising several camera viewpoints (see Goffredo et al., 2010).

The most widely implemented method of gait analysis involves integration with video surveillance (Chaurasia et al., 2015). This facilitates automatic analysis of routinely collected surveillance footage (Zhang et al., 2011). This is significant, because gait analysis cannot only be used to identify individuals, but also to identify and automatically alert police and security to abnormal movement and behaviour (Zhang et al., 2011). Gait analysis can also be used to identify individuals and track their movement through public spaces in real-time (Zhang et al., 2011).

There are several different types of gait analysis and recognition, depending on the specific technology used (De Marsico & Mecca, 2016). In addition to vision-based approaches, floor-based technology can capture walking pattern and weight, where specialist equipment has been installed. Gait recognition can be conducted through the use of wearable sensors used to capture baseline data about the way an individual moves (Zhang et al., 2011). Current developments in the area of gait recognition includes the use of infrared image sequences from video footage taken at night (Lee, Belkhatir & Sanei, 2014), and identification on the basis of the sound of footsteps, known as acoustic analysis (Altaf, Butko & Juang, 2015).

Application and issues

The fact that gait analysis can be conducted from a distance means that a key application of this biometric is integration with video surveillance systems. Gait analysis can enable automated identification detection at a distance, in contrast with most other biometric modalities (Lee et al., 2014). Other applications include automatic door opening in security-sensitive environments such as banks and airports (Indumathi & Pushparani, 2016). Makihara et al. (2015) discuss the application of gait analysis to secure a criminal conviction in a 2004 bank robbery case in Denmark, solved using gait analysis comparison of video footage from the crime scene and subsequent recordings of suspects.

Gait recognition differs from other biometric modalities in a number of important ways. As has been discussed, gait recognition and face recognition are the only biometric modalities that can currently be used to identify and monitor or track people unobtrusively without their cooperation or knowledge (Katiyar, Pathak & Arya, 2013). Gait recognition can occur at distances of 10 metres or more (Lee et al., 2014). In contrast, facial recognition is more suitable at short range, and requires higher-resolution images (Lee et al., 2014). If a complete facial image cannot be obtained, or an individual hides their face, or distance and image quality are unsuitable for facial recognition, gait recognition may provide a suitable alternative. The combination of gait recognition and facial recognition can enable improved accuracy (Zhang et al., 2011).

One concern with gait recognition is the potential for learned replication of an individual's walking style to defeat the recognition system (Hadid et al., 2015). Recent research has attempted to devise new methods of analysis to overcome these issues; however, in general, gait recognition can provide an important additional layer of security when used in combination with other modalities, such as face or footprint biometrics (Chaurasia et al., 2015; Katiyar et al., 2013).

Keystroke dynamics

Principles

Keystroke dynamic recognition enables authentication via the identification of individual typing characteristics and patterns, including key press durations (Revett, 2009). Although keystroke dynamic recognition was first invented in the 1980s, it is now being used more frequently, in line with the increased use of computers and the expansion of the Internet (Rudrapal, Das & Debbarma, 2014). Like other forms of behavioural biometrics, keystroke dynamics are generally considered less reliable than physiological biometrics due to the variability of this type of human behaviour (Revett, 2009).

At the enrolment stage of keystroke dynamic recognition, individual typing characteristics are extracted to create a digital typing signature (Revett, 2009). At enrolment the user is typically asked to repeatedly enter their details to extract the typing profile (Revett, 2009). These characteristics are used to develop a profile of an individual user that forms a reference for future verification (Revett, 2009). However, some researchers have argued that keystroke latency and duration is not sufficient for authentication, and proposed other combinations of typing characteristic metrics (Rudrapal et al., 2014; Ngugi, Tarasewich & Reece, 2012). A combination of different metrics results in higher authentication accuracy (Ngugi et al., 2012).

Keystroke dynamic recognition is less accurate than other forms of biometric recognition; however, it is difficult to compare accuracy rates for keystroke dynamic recognition across the literature, as different studies use a variety of metrics. Reliability is directly related to the length of text typed. For example, in a study by Bergadano, Gunetti & Picardi (2002) a false negative rate of 4 per cent and a false positive rate of less than 0.01 per cent were obtained. However, in this study, the participants were required to type 683 characters, a length that would be too long for a password, and may inhibit wide-scale adoption, depending on the context. If keystroke dynamics are used for short passwords, this raises questions about the accuracy of the authentication (Ngugi et al., 2012).

Application and issues

With increasing threats to computer systems and information security, keystroke dynamics could play a key role in strengthening computer security. There are a range of applications for keystroke dynamic recognition, including providing stronger authentication, identity confirmation, user identification and tracking over the Internet (see discussion in Bergadano et al., 2002). Keystroke dynamics can enhance computer security by adding an additional layer of authentication in addition to passwords. Keystroke dynamic recognition is used in this way to strengthen passwords, but it is not typically used alone as a single factor for authentication, due to the issues of accuracy and variability that have been raised (Rudrapal et al., 2014).

Keystroke recognition can be static, occurring at login, or continuous, as a person is typing and interacting with a computer (Monrose & Rubin, 2000). Software has been developed for use in academic settings to continuously monitor student typing to help prevent plagiarism. Keystroke dynamic recognition can also be conducted over the Internet, opening possibilities for remote authentication. Further applications for keystroke dynamic recognition are the use of behavioural biometric keypads with pressure sensors that are integrated with access points or automatic teller machines. Current research is applying keystroke recognition features to smart phone touchscreens (Kambourakis et al., 2016).

Some of the advantages of keystroke dynamic recognition include that it is software-based, unobtrusive, can be conducted over the Internet and has a low implementation cost (Ngugi et al., 2012). Users are already familiar with authentication of their identity with logins and passwords, and, from this perspective, it may be one of the more acceptable forms of biometrics (Revett, 2009; Karnan, Akila & Krishnaraj, 2011). It is expected that further use of keystroke dynamic recognition will occur with increased accuracy in authentication aligned with the uptake and development of pressure-sensitive keyboards that have recently been developed and widely marketed (Ngugi et al., 2012).

Cognitive biometrics

Principles

Cognitive biometrics are defined as 'methods and technologies for the recognition of humans, based on the measurement of signals generated directly or indirectly from their thought processes' (Revett, Deravi & Sirlantzis, 2010, p. 71). These systems establish authentication via biosignals that reflect the mental states of individuals, as measured by brain-computer interfaces (BCIs) (Jolfaei, Wu & Muthukkumarasamy, 2013). The use of cognitive biometric systems has become the subject of increasing attention as the technology has continued to develop in recent years (Jolfaei et al., 2013; Armstrong et al., 2015).

Neural activity can be used as a biometric signature that reflects individual mental activities or cognitive processes (Tsuru & Pfurtscheller, 2012). Cognitive biometrics involve the use of an electroencephalogram (EEG) which is non-invasive and captures electrical signals produced by the firing of neurons within the brain; it is used in medicine to measure brain function. This can be undertaken when an individual performs a certain cognitive task, such as visual perception, memory or language tasks that activate specific regions of the brain and lead to specific patterns in EEG activity (Revett et al., 2010). When electrical signals are associated with a specific stimulus, an event-related potential (ERP) can be obtained (Armstrong et al., 2015). ERPs therefore correspond to specific cognitive events, for example, thinking of a specific password (Armstrong et al., 2015). Empirical evidence indicates that humans 'generate recordable and reproducible signals that can be captured using EEG technology when we think of something as a password' (Revett et al., 2010, p. 74). Instead of using a password humans may be able to authenticate by simply thinking of a specific thing, or password (Revett et al., 2010). The use of the EEG has provided promising results in classification accuracy approaching physiological-based approaches such as fingerprinting (Revett et al., 2010). Armstrong et al. (2015) were able to label ERPs as belonging to specific individuals with an accuracy rate that ranges from 82 to 97 per cent, and found to be stable over time, using a technique that requires three electrodes to be placed on the scalp.

Application and issues

As ERPs can be used as cognitive passwords, there are possible applications in user identification, however cognitive biometrics are not currently widely adopted. Cognitive biometrics are considered to be highly resistant to circumvention (Revett et al., 2010). Research supports the proposition that an individual's brain-wave patterns are unique and 'nearly impossible to forge or duplicate as the neural activity of people are distinctive even when they think about the same thing' (Bajwa & Dantu, 2016, p. 95).

The main disadvantage of cognitive biometrics is that they require the use of sensitive EEG equipment, including electrodes and conductive gels, and, for this reason, their use in everyday settings may not be realistic (Revett et al., 2010). The use of EEG equipment is currently prohibitively expensive on a wide scale, but this may change in accordance with the development of new technology (Bajwa & Dantu, 2016).

References

Abaza, A., Ross, A., Hebert, C., Harrison, M. & Nixon, M. (2013). A survey on ear biometrics. *ACM Computing Surveys 45*(2), 1.

Ali, A. & Islam, M. (2013). A biometric based 3D ear recognition system combining local and holistic features. *International Journal of Modern Education and Computer Science 11*, 36.

Altaf, M., Butko, T. & Juang, B. (2015). Acoustic gaits: Gait analysis with footstep sounds. *IEEE Transactions on Biomedical Engineering 62*(8), 2001.

Anwar, A., Ghany, K. & Elmahdy, H. (2015). Human ear recognition using geometrical features extraction. *Procedia Computer Science 65*, 529.

Armstrong, B., Ruiz-Blondet, M., Kahalifian, N., Kurtz, K., Jun, Z. & Laszlo, S. (2015). Brainprint: Assessing the uniqueness, collectability, and permanence of a novel method for ERP biometrics. *Neurocomputing 166*, 59.

Australian Taxation Office. (2014, 8 September). ATO launches voice authentication: Australians can save time on the phone to the ATO. Retrieved from https://www.ato.gov.au/media-centre/media-releases/ato-launches-voice-authentication

Bajwa, G. & Dantu, R. (2016). Neurokey: Towards a new paradigm of concealable biometrics-based key generation using electroencephalograms. *Computers and Security 62*, 95.

Barra, S., Casanova, A., Narducci, F. & Ricciardi, S. (2015). Ubiquitous iris recognition by means of mobile devices. *Pattern Recognition Letters 57*, 66.

Benziane, S. & Benyettou, A. (2016). Dorsal hand vein identification. *International Journal of Computer Science and Information Security 14*(3), 423.

Bergadano, F., Gunetti, D. & Picardi, C. (2002). User authentication through keystroke dynamics. *ACM Transactions on Information and System Security 5*(4), 367.

Borgen, H., Bours, P. & Wolthusen, S. (2009). Simulating the influences of ageing and ocular disease on biometric recognition performance. In Tistarelli, M. & Nixon, M.S. (eds.), *Advances in Biometrics: Third International Conference* (pp. 857–867). Alghero, Italy: Springer.

Bouchrika, I., Carter, J. & Nixon, M. (2016). Towards an automated visual surveillance using gait for identity recognition and tracking across multiple non-intersecting cameras. *Multimedia Tools and Applications 75*, 1210.

Bowyer, K., Ortiz, E., & Sgroi, A. (2015). Iris recognition technology evaluated for voter registration in Somaliland. *Biometric Technology Today 2*, 5.

Chaurasia, P., Yogarajah, P., Condell, J., Prasad, G., McIlhatton, D. & Monaghan, R. (2015). Biometrics and counter-terrorism: The case of gait recognition. *Behavioural Sciences of Terrorism and Political Aggression* 7(3), 210.

Chenafa, M., Istrate, D., Vrabie, V. & Herbin, M. (2008). Biometric system based on voice recognition using multiclassifiers. In Schouten, B., Juul, N.C., Drygaijlo, A. & Tistarelli, M. (eds.), *Biometrics and Identity Management: First European Workshop* (pp. 206–215). Roskilde, Denmark: Springer.

Choudhury, S. & Tjahjadi, T. (2013). Gait recognition based on shape and motion analysis of silhouette contours. *Computer Vision and Image Understanding 11*, 1770.

Collinson, P. (2014, 14 May). Forget fingerprints – banks are starting to use vein patterns for ATMs. *The Guardian*. Retrieved from https://www.theguardian.com/money/2014/may/14/fingerprints-vein-pattern-scan-atm

Colores, J., Garcia-Vazquez, M., Ramirez-Acosta, A. & Perez-Meana, H. (2011). Iris image evaluation for non-cooperative biometric iris recognition systems. In Batyrshin, I. & Sidorov, G. (eds), *Advances in Soft Computing: 10th Mexican International Conference on Artificial Intelligence*. Puebla, Mexico: Springer.

De Marsico, M. & Mecca, A. (2016). Biometric walk recognizer: Gait recognition by a single smartphone accelerometer. *Multimedia Tools and Applications 75, 1201.* doi:10.1007/s11042-11014-2364-2369.

Galka, J., Masior, M. & Salasa, M. (2014). Voice authentication embedded solution for secured access control. *IEEE Transactions on Consumer Electronics 60*(4), 653.

Goffredo, M., Bouchrika, I., Carter, J. & Nixon, M. (2010). Performance analysis for automated gait extraction and recognition in multi-camera surveillance. *Multimedia Tools and Applications 50*, 75.

Hadid, A., Ghahramani, M., Kellokumpu, V., Feng, X., Bustard, J. & Nixon, M. (2015). Gait biometrics under spoofing attacks: An experimental investigation. *Journal of Electronic Imaging 24*(6), 1.

Indumathi, T. & Pushparani, M. (2016). Automatic door opening using gait identification for movement as gesture. *Journal of Engineering Technology 4*(1), 132.

Ives, R.Broussard, R., Rakvic, R. & Link, S. (2013). Iris recognition. In Du, E. (ed.), *Biometrics: From Fiction to Practice* (pp. 65–86). Boca Raton, FL: Pan Stanford Publishing.

Jolfaei, A., Wu, X. & Muthukkumarasamy, V. (2013). On the feasibility and performance of pass-thought authentication systems. In McDonald-Maier, K.D., Howells, G. & Stoica,

A. (eds), *IEEE Computer Society 2013 Fourth International Conference on Emerging Security Technologies* (pp. 33–38). Cambridge: Conference Publishing Services.

Kaman, S., Swetha, K., Akram, S. & Varaprasad, G. (2013). Remote user authentication using a voice authentication system. *Information Security Journal: A Global Journal 22*(3), 117.

Kambourakis, G., Damopoulos, D., Papamartzivanos, D. & Pavlidakis, E. (2016). Introducing touchstroke: Keystroke-based authentication systems for smartphones. *Security and Communication Networks 9*, 542.

Karnan, M., Akila, M. & Krishnaraj, N. (2011). Biometric personal authentication using keystroke dynamics: A review. *Applied Soft Computing 11*, 1565.

Katiyar, R., Pathak, V. & Arya, K. (2013). A study on existing gait biometric approaches and challenges. *International Journal of Computer Science Issues 10*(1), 135.

Kumar, V.K. & Srinivasan, B. (2014). Automated human identification scheme using ear biometric technology. *International Journal of Image, Graphics and Signal Processing 3*, 58.

Latman, N. & Herb, E. (2013). A field study of the accuracy and reliability of a biometric iris recognition system. *Science and Justice 53*, 98.

Lecher, C. & Brandom, R. (2016, July 12). The FBI has collected 430,000 iris scans in a so-called 'pilot program'. *The Verge*. Retrieved from http://www.theverge.com/2016/7/12/12148044/fbi-iris-pilot-program-ngi-biometric-database-aclu-privacy-act

Lee, T., Belkhatir, M. & Sanei, S. (2014). A comprehensive review of past and present vision-based techniques for gait recognition. *Multimedia Tools and Applications 72*, 2834.

Makihara, Y., Yahi, Y., Matovski, D.S., Nixon, M.S. & Carter, J.N. (2015). Gait recognition: Databases, representations, and applications. In Webster, J. (ed.), *Wiley Encyclopedia of Electrical and Electronics Engineering* (pp. 1–15). Chichester: John Wiley and Sons, Inc.

Mazaira-Fernandez, L., Alvarez-Marquina, A. & Gomez-Vilda, P. (2015). Improving speaker recognition by biometric voice deconstruction. *Frontiers in Bioengineering and Biotechnology 3*, 1.

Meyer, D. (2017). Ultrasecure Samsung Galaxy S8 iris scanner can be easily tricked, say hackers. *Znet*. Retrieved from http://www.zdnet.com/article/ultrasecure-samsung-galaxy-s8-iris-scanner-can-be-easily-tricked-say-hackers

Monrose, F. & Rubin, A. (2000). Keystroke dynamics as a biometric for authentication. *Future Generation Computer Systems 16*, 351.

Morgen, B. (2012). Voice biometrics for customer authentication. *Biometric Technology Today*, February, 8.

Neves, J., Narducci, F., Barra, S. & Proenca, H. (2016). Biometric recognition in surveillance scenarios: A survey. *Artificial Intelligence Review 46*, 515.

Ngugi, B., Tarasewich, P. & Reece, M. (2012). Typing biometric keypads: Combining keystroke time and pressure features to improve authentication. *Journal of Organizational and End User Computing 24*(1), 42.

Nigam, I., Vatsa, M. & Singh, R. (2015). Ocular biometrics: A survey of modalities and fusion approaches. *Information Fusion 26*, 1.

Nixon, M. & Carter, J. (2006). Automatic recognition by gait. *Proceedings of the Electrical and Electronics Engineers 94*(11), 2013.

Nuance Communications. (2017). Voice biometrics. Retrieved from http://australia.nuance.com/for-business/customer-service-solutions/voice-biometrics/index.htm

Pflug, A. & Busch, C. (2012). Ear biometrics: A survey of detection, feature extraction and recognition methods. *Institution of Engineering and Technology Biometrics 1*(2), 114.

Pierscionek, B., Crawford, S. & Scotney, B. (2008). Iris recognition and ocular biometrics: The salient features. In Scotney, B. & Morrow, P. (eds), *International Machine Vision and Image Processing Conference* (pp. 170–175). Los Alamitos, CA: IEEE Computer Society.

Pun, K. & Moon, Y. (2004). Recent advances in ear biometrics. In Azada, D. (ed.), *Proceedings of the Sixth IEEE International Conference on Automatic Face and Gesture Recognition* (pp. 164–169). Seoul, South Korea: IEEE Computer Society.

Revett, K. (2009). A bioinformatics based approach to user authentication via keystroke dynamics. *International Journal of Control, Automation and Systems* 7(1), 7.

Revett, K., Deravi, F. & Sirlantzis, K. (2010). Biosignals for user identification: Towards cognitive biometrics? In Howells, G., Sirlantzis, K., Stoica, A., Huntsberger, T. & Arslan, T. (eds), *IEEE Computer Society 2010 Conference on Emerging Security Technologies* (pp. 71–76). Canterbury: Conference Publishing Services.

Rudrapal, D., Das, S. & Debbarma, S. (2014). Improvisation of biometrics authentication and identification through keystroke pattern analysis. In Natarajan, R. (ed.), *Distributed Computing and Internet Technology: 10th International Conference* (pp. 287–292). Bhubaneswar, India: Springer.

Sandhu, P. & Kaur, M. (2015). A survey: Multimodal systems of finger vein and iris. *International Journal of Computer Science and Information Security 13*(3), 79.

Smorawa, D. & Kubanek, M. (2015). Biometric systems based on palm vein patterns. *Journal of Telecommunications and Information Technology 2*, 18.

Tsuru, K. & Pfurtscheller, G. (2012). Brainwave biometrics: A new feature extraction approach with cepstral analysis method. *Japanese Society for Medical and Biological Engineering 50*(1), 162.

Waheed, Z., Akram, M., Waheed, A., Khan, M., Shaukat, A. & Ishaq, M. (2016). Person identification using vascular and non-vascular retinal features. *Computers and Electrical Engineering 53*, 359.

Wang, Y., He, D., Yu, Chhongchong, Y., Tongqiang, J. & Zaiwen, L. (2012). Multimodal biometrics approach using face and ear recognition to overcome adverse effects of pose changes. *Journal of Electronic Imaging 21*(4), 1.

Wu, H., Wang, Y. & Huang, J. (2014). Identification of electronic disguised voices. *IEEE Transactions on Information Forensics and Security 9*(3), 489.

Zhang, Z., Hu, M. & Wang, Y. (2011). A survey of advances in biometric gait recognition. In Sun, Z., Lai, J., Chen, X. & Tan, T. (eds), *Biometric Recognition: 6th Chinese Conference* (pp. 150–158). Beijing, China: Springer.

6

BIOMETRICS IN CRIMINAL TRIALS

Introduction

This chapter explores the ways in which criminal courts have dealt with the emergence of biometrics as a source of evidence. The main purpose in considering this kind of evidence in criminal trials and appeals is to establish the identity of either the offender or the victim. As discussed in previous chapters, fingerprint and DNA analysis have long been admitted as evidence aiding identification, and these have been supplemented more recently by facial and body mapping, voice analysis and other types of biometrics. However, each of these has faced challenges to acceptance as a form of evidence, based mainly on concerns about their reliability, regulatory control and the manner of their presentation in legal proceedings. This chapter provides an insight into the trend of accepting biometric identification as a source of evidence, but with some judicial reservations about the application of particular kinds of biometrics in the criminal justice system. These concerns are largely based on whether certain forms of identification have achieved the degree of scientific reliability that is required for legal admissibility.

Identification evidence

Before discussing the main forms of biometrics that courts deal with, it is useful to consider the context of such evidence.[1] In criminal trials, the prosecution is

1 In this chapter, the focus is on criminal proceedings. However, biometrics can also play a role in civil or administrative proceedings. An example is the resolution of paternity claims in family law: see, for example, the case of *Magill v Magill* [2006] HCA 51; (2006) 231 ALR 277; (2006) 81 ALJR 254 (9 November 2006) in which DNA testing after the end of a marriage revealed that two children were not biological offspring of the father, leading to tortious claims of deceit. Biometrics are also used in migration cases to

required to prove its case against the defendant (also referred to as the accused) beyond reasonable doubt, unless there is a guilty plea and the matter proceeds directly to sentencing. Where the prosecution is required to prove the identity of the person who allegedly committed the crime, there will usually be some form of 'identification evidence' adduced. An example of this form of evidence is the following:[2]

> **"identification evidence"** means evidence that is:
>
> (a) an assertion by a person to the effect that a defendant was, or resembles (visually, aurally or otherwise) a person who was, present at or near a place where:
>
> (i) the offence for which the defendant is being prosecuted was committed; or
>
> (ii) an act connected to that offence was done;
>
> at or about the time at which the offence was committed or the act was done, being an assertion that is based wholly or partly on what the person making the assertion saw, heard or otherwise perceived at that place and time; or
>
> (b) a report (whether oral or in writing) of such an assertion.

In most cases, the assertion of identity will be made by a person who was an 'eye witness' at the scene of a crime, and has made such a report to police or is able to do so in court testimony.[3] The provisions dealing with identification evidence impose as a general pre-condition to the admissibility of such evidence that the defendant participated in an 'identification parade' or, as it is also known, a 'police line-up'.[4] This requirement is due to the fact that eye witness identification has historically been seen as unreliable and has led in some instances to wrongful convictions, so that more controlled and supervised identification procedures are preferred.[5]

help in establishing or verifying identity: see, for example, *SZSZM v Secretary, Department of Immigration and Border Protection* [2017] FCA 458 (27 April 2017).

2 This example is from the uniform evidence law (UEL) that operates in several Australian jurisdictions.

3 The expression 'visually, aurally or otherwise' allows other senses to form the basis of a witness identification, such as recognising a distinctive voice, accent, posture or gait. A recent case in which a 'voice identification parade' was used is *Miller v R* [2015] NSWCCA 206 (3 August 2015).

4 Section 113 provides that the identification evidence provisions only apply in criminal proceedings. Sections 114 and 115 refer to the use of an 'identification parade' but do not define the term. Section 114 deals with 'visual identification evidence' while s115 deals with 'picture identification evidence'. The conduct of identification parades is governed by other legislation such as the *Crimes Act 1914* (Cth), ss3ZM and 3ZN. Section 116 imposes requirements for warnings to the jury in relation to identification evidence.

5 The High Court of Australia summarised the problems with eye witness identification almost a century ago, and noted requirements for identification procedures that are still

Biometrics provide an alternative to traditional eye witness identification. Not quite falling within the definition of 'identification evidence' set out above, biometric forms of identification may nonetheless be admissible as a type of circumstantial evidence.[6] Almost invariably, such evidence is presented in the form of expert opinion evidence, either through a forensic analyst report or by means of witness testimony from the specialist who conducted biometric analysis.

General requirements for admissibility of biometric identification are that the evidence must be:

a ***relevant*** in the proceeding, meaning that it has the capacity to help resolve factual issues in the trial, such as the identity of an offender;[7]

b based on ***specialised knowledge***, meaning that it must be presented by an expert who has previous training, study or experience in the applicable field of expertise;[8]

c not ***unfairly prejudicial***, in which case it may be ruled inadmissible by the judge.[9]

Techniques of biometric identification that have been considered by courts include facial and body mapping, fingerprinting and DNA matching. These are discussed in

followed today: *Davies (and Cody) v The King* [1937] HCA 27; (1937) 57 CLR 170; see also *Alexander v The Queen* [1981] HCA 17; (1981) 145 CLR 395.

6 Biometric identification such as a fingerprint or DNA match is not classed as 'identification evidence' because it is usually not based on what 'the person making the assertion saw, heard or otherwise perceived at that place and time' but on later forensic analysis by a person who was not a witness to the events in question: see Australian Law Reform Commission, *Uniform Evidence Law* (ALRC 102), [13.25]. This means that Part 3.9 does not apply, and biometric evidence is treated as a form of circumstantial evidence; for example, the judge in *R v Pfennig (No. 2)* [2016] SASC 171 (11 November 2016), [31] stated: 'I point out, however, that the DNA evidence is not direct evidence going to the guilt of the accused. I treat it as circumstantial evidence to be considered alongside all of the other evidence in the case'.

7 Section 55(1) of the UEL legislation provides: 'The evidence that is relevant in a proceeding is evidence that, if it were accepted, could rationally affect (directly or indirectly) the assessment of the probability of the existence of a fact in issue in the proceeding'. Relevant evidence is admissible subject to other provisions: s56.

8 Section 79(1) provides an exception from the exclusionary opinion rule in s76 as follows: 'If a person has specialised knowledge based on the person's training, study or experience, the opinion rule does not apply to evidence of an opinion of that person that is wholly or substantially based on that knowledge'. An expert may give this evidence in the form of affidavit under s177, or may be called to give the evidence through testimony.

9 In particular, s137 provides: 'In a criminal proceeding, the court must refuse to admit evidence adduced by the prosecutor if its probative value is outweighed by the danger of unfair prejudice to the defendant'. Additionally, s138(1) provides: 'Evidence that was obtained: (a) improperly or in contravention of an Australian law; or (b) in consequence of an impropriety or of a contravention of an Australian law; is not to be admitted unless the desirability of admitting the evidence outweighs the undesirability of admitting evidence that has been obtained in the way in which the evidence was obtained'. Thus, investigative practices may also affect admissibility.

turn below. However, as discussed in Chapter 5, new techniques are always evolving and this list is not exhaustive.

Facial and body mapping

The biometrics of facial mapping and body mapping primarily involve the comparison of still images in order to determine likely identity, typically between an image taken from a crime scene and a comparable image depicting a criminal defendant (discussed in Chapter 4). For example, photographs developed from a CCTV camera recording can be compared with photographs of the defendant, or even directly with the defendant's appearance in the courtroom. This kind of visual comparison can in some instances be made by a jury without assistance, and indeed courts have sometimes stressed that to allow police or other witnesses to offer their opinions of similarity as evidence can usurp the proper role of the jury.[10] However, facial and body mapping techniques usually involve some level of technical skill and measurement that goes beyond what a lay jury can do, thus making it properly the subject of expert evidence. These techniques are closest to being a scientific analogue of traditional eye witness identification, which is notoriously susceptible to inaccuracy. However, the scientific reliability of facial and body mapping has been questioned, including in criminal proceedings.

Significant cases

A noteworthy early case involving this form of evidence that was widely publicised in the United Kingdom and Australia was the murder trial arising from the disappearance in the Northern Territory of British tourist Peter Falconio, whose body was never found (Gans, 2007c). Part of the evidence was a photographic image developed from security camera footage at a highway truck stop, which was compared by a facial mapping expert called by the prosecution with images of the defendant. This evidence was allowed by the trial judge in the case, along with DNA evidence linking the defendant to the crime:[11]

10 *Smith v The Queen* [2001] HCA 50; (2001) 206 CLR 650, in which a High Court majority observed: 'Because the witness's assertion of identity was founded on material no different from the material available to the jury from its own observation, the witness's assertion that he recognised the appellant is not evidence that could rationally affect the assessment by the jury of the question we have identified. The fact that someone else has reached a conclusion about the identity of the accused and the person in the picture does not provide any logical basis for affecting the jury's assessment of the probability of the existence of that fact when the conclusion is based only on material that is not different in any substantial way from what is available to the jury.'

11 *The Queen v Murdoch* [2005] NTSC 78 (15 December 2005), (Martin CJ), [207]-[208]. DNA aspects of the case are discussed in two articles by Jeremy Gans, 'The Peter Falconio Investigation: Needles, Hay and DNA' (2007c) and 'Catching Bradley Murdoch: Tweezers, Pitchforks and the Limits of DNA Sampling' (2007a).

> The image of the person entering the shop at the truck shop taken from the security film is far from clear. This is not a case of comparing clear photographs where it could be said with considerable force that the jury could reach its own conclusion without help. In addition, there is evidence that the accused has changed his appearance since July 2001. The comparison between the image from the security film and photographs of the accused is far from straightforward and, in my opinion, the jury would be assisted by the evidence of Dr Sutisno.
>
> Further, in my view, it is not appropriate to limit the assistance to merely identifying the relevant characteristics. When regard is had to the nature and detail of the characteristics and the methodology employed by Dr Sutisno, it is readily apparent that her knowledge and expertise in the area of anatomy give Dr Sutisno a significant advantage in the assessment of the significance of the features of comparison both individually and in their combination. Dr Sutisno possesses scientific knowledge, expertise and experience outside the ordinary knowledge, expertise and experience of the jury. This is not a case in which the jury, having been informed of the relevant features, would not be assisted by the expert evidence of Dr Sutisno as to her opinion of the significance of the features individually and in their combination.

The court was also prepared to accept body mapping, a more recent technique involving superimposition of images, as an extension of facial mapping:[12]

> Body mapping has received limited attention within the scientific community. For that reason it may be regarded as a new technique, but as Dr Sutisno explained it is merely an extension of the well recognised and accepted principles of facial mapping to the remainder of the body. I am satisfied that the technique has 'a sufficient scientific basis to render results arrived at by that means part of a field of knowledge which is a proper subject of expert evidence'.

However, on appeal it was held that the facial and body mapping evidence should not have been allowed beyond the expert assisting the jury to ascertain physical similarities, rather than in the expert reaching conclusions about identity:[13]

> This Court has found that the technique employed by Dr Sutisno did not have a sufficient scientific basis to render the results arrived at by that means part of a field of knowledge which is a proper subject of expert evidence. However the evidence given by Dr Sutisno was capable of assisting the jury in

12 *The Queen v Murdoch* [2005] NTSC 78 (15 December 2005), [110].

13 *Murdoch v The Queen* [2007] NTCCA 1 (10 January 2007), [300]. Despite this ruling, however, the conviction was upheld as it was amply supported by other evidence. A special leave application to the High Court was unsuccessful: *Murdoch v The Queen* [2007] HCATrans 321 (21 June 2007).

> terms of similarities between the person depicted in the truck stop footage and the appellant. It was evidence that related to, and was admissible as, demonstrating similarities but was not admissible as to positive identity. Dr Sutisno was not qualified to give evidence, as she did, based on "face and body mapping" as to whether the two men were, indeed, the same man. Her evidence in this regard should not have been received.

Facial and body mapping evidence can therefore be admitted in criminal proceedings, but its use must be managed so as not to usurp the function of the jury as decider of the facts. Two other cases decided at around the same time reached a similar conclusion, though with some additional differentiation between facial and body mapping. In the case of *Tang*, the Court of Criminal Appeal considered the expert's use of biometric methods:[14]

> Dr Sutisno compared measurements and dimensions of faces (photo-anthropometry) and individual facial and body features (morphological analysis). She magnified photographs of the offender to the same size as the suspect, changed the opacity of one before putting it on top of the other, in order to see whether the features aligned or one could be overlayed over the other (photograph superimposition). Furthermore, she identified distinctive individual characteristic and habits, which she called "unique identifiers".
>
> Dr Sutisno used photo-anthropometry as a first step in facial and body mapping, but did not rely solely on the findings of this procedure because of the possibility of two or more people having the same dimensions. She regards morphological analysis as more accurate than photo-anthropometry, because it compares individual facial and body features and takes into account distinctive characteristic habits of the individual. She asserts that morphological analysis provides results sufficient to show whether two sets of photographs [of] people were of the same person or not.

The court went on to consider similarities between these techniques and other biometrics such as fingerprint comparison, a more established method of forensic identification. By analogy, it was accepted that expert evidence of similarities, derived from comparison of facial or body photographs could also provide assistance to the jury, including acquired or 'ad hoc' expertise:[15]

14 *R v Hien Puoc Tang* [2006] NSWCCA 167 (24 May 2006), [19]-[20] (Spigelman CJ).

15 *R v Hien Puoc Tang* [2006] NSWCCA 167 (24 May 2006), [120] (Spigelman CJ). The concept of *ad hoc* expertise in relation to voice identification has been applied in cases such as *Butera v Director of Public Prosecutions (Vic)* [1987] HCA 58; (1987) 164 CLR 180; *R v Leung and Wong* [1999] NSWCCA 287; and more recently, *Morgan v R* [2016] NSWCCA 25 (26 February 2016); and *Nasrallah v R; R v Nasrallah* [2015] NSWCCA 188 (17 July 2015).

> The process of identification and magnification of stills from the videotape was a process that had to be conducted by Dr Sutisno out of court. Furthermore, the quality of the photographs derived from the videotape was such that the comparison of those stills with the photographs of the Appellant could not be left for the jury to undertake for itself. The identification of points of similarity by Dr Sutisno was based on her skill and training, particularly with respect to facial anatomy. It was also based on her experience with conducting such comparisons on a number of other occasions. Indeed, it could be supported by the experience gained with respect to the videotape itself through the course of multiple viewing, detailed selection, identification and magnification of images. By this process she had become what is sometimes referred to as an "ad hoc expert".

However, in order for the opinions of identity offered by the expert in this case to be admissible, compliance with the specialised knowledge requirements of evidence law had to be demonstrated. The court ruled that there was an inadequate connection between the body mapping techniques being applied and the 'training, study or experience' of the expert, and thus the opinions on offer did not pass the requirements of the relevant evidence law:[16]

> In the case of the Appellant the relevant evidence about posture was expressed in terms of "upright posture of the upper torso" or similar words. The only links to any form of "training, study or experience" was the witnesses' study of anatomy and some experience, entirely unspecified in terms of quality or extent, in comparing photographs for the purpose of comparing "posture". The evidence in this trial did not disclose, and did not permit a finding, that Dr Sutisno's evidence was based on a study of anatomy. That evidence barely, if at all, rose above a subjective belief and it did not, in my opinion, manifest anything of a "specialised" character. It was not, in my opinion, shown to be "specialised knowledge" within the meaning of s79.

In the *Jung* case a month later, a judge was again required to rule on the admissibility of Dr Sutisno's facial and body mapping analysis in a murder trial. In this case, the defence called its own expert witnesses, who cast doubt on the claims made for the techniques, referring to the quality of the photographs used. Nonetheless, the judge ruled the evidence admissible, with questions of the quality of the analysis going to its weight rather than admissibility:[17]

> However adequate or inadequate the photographic materials utilised by Sutisno for the purpose of her analysis, the evidence on the voir dire does not

16 *R v Hien Puoc Tang* [2006] NSWCCA 167 (24 May 2006), [140] (Spigelman CJ, Simpson and Adams JJ agreeing).
17 *R v Jung* [2006] NSWSC 658 (29 June 2006), [62]-[64].

> establish that she has failed to disclose the factual material she has utilised (the photographic images), the nature of the methodology that she has employed and the type of analysis described in her reports (morphological analysis). I have carefully reviewed the reports and her evidence in order to determine whether it may properly be said that, having regard to the specific principles governing admissibility of expert evidence as identified by Heydon, JA in *Makita* … Dr. Sutisno's evidence complies with the requirements for admissibility.
>
> Insofar as she has identified the relevant factual matters that she has taken into account (the particular photographic images) the particular facial features which she maintains are examinable by reference to such images and the nature of the methodology employed by her, the tests of admissibility in those respects are satisfied. The question of the weight, including the reliability, of the opinion is, of course, a quite different matter and it is anticipated at trial that attention will be given to the quality of the photographic images, their alleged deficiencies and the significance that arises from those matters.

Another case to examine the scientific reliability of the technique of body mapping was based on a comparison, of both moving and still images of an offender and the defendant, by an expert in the field of anthropology and comparative anatomy. In overturning the conviction in this case on appeal, the court expressed its concern about the 'lack of research into the validity, reliability and error rate of the process'.[18] Thus, the scientific reliability of body mapping has not been definitively resolved.

Three years later in 2014, similar evidence was considered in the *Honeysett* case (discussed in Buckland, 2014; Edmond & San Roque, 2014). The opinion of the expert, based on body mapping analysis in an armed robbery case, identified the appellant:[19]

> He is an adult male of ectomorphic (thin, 'skinny') body build. His shoulders are approximately the same width as his hips. His body height is medium compared to other persons, and to familiar objects (eg doorways) visible in the images from the [offence]. He carries himself very straight, so that his hips are standing forward while his back has a very clearly visible lumbar lordosis (the small of his back is bent forward) overhung by the shoulder area. Although the offender covers his head and face with a cloth (what looks like a T-shirt) … the knitted fabric is elastic and adheres closely to the vault of his skull (= braincase). This shows that his hair is short and does not distort the layout of the fabric. The shape of the head is clearly dolichocephalic (= long head, elongated oval when viewed from the top) as opposed to brachycephalic (= short head, nearly spherical). The offender is right-handed in his actions …

18 *Morgan v R* [2011] NSWCCA 257 (1 December 2011), [138] (Hidden J).
19 *Honeysett v The Queen* [2014] HCA 29 (13 August 2014), [14]-[17].

> Although most of the body of the offender is covered by clothing, head wrap and gloves, an area of naked skin above his wrist (between the glove and the sleeve) in images … is visible and can be compared to the skin colour of a female hotel employee on the same images.
>
> [The appellant] is an adult male of ectomorphic (= slim) body buil[d]. His hips and shoulders are of approximately the same width. His stance is very straight with well marked lumbar lordosis and pelvis shifted forward. His skull vault is dolichocephalic when viewed from the top. Comparison of lateral (side) and front views of his head also indicates the head … is long but narrow. His skin is dark, darker than that of persons of European extraction, but not 'black' … He is right-handed – uses his right hand to sign documents.

The expert concluded that there was a 'high degree of anatomical similarity' between the offender and the appellant, and this opinion was 'strengthened by the fact that he was unable to discern any anatomical dissimilarity between the two individuals'. This evidence was allowed to be heard by the jury, which convicted the appellant. On appeal, the court held that the evidence fell within the 'training, study or experience', of the expert witness.[20] The appeal was dismissed and the matter went before a High Court for further consideration.

The High Court unanimously agreed that, whatever the scientific merits of body mapping as a reliable and validated field of study, the expert's opinion in this case was simply not sufficiently based on his expertise in anatomy:[21]

> Professor Henneberg's opinion was not based on his undoubted knowledge of anatomy. Professor Henneberg's knowledge as an anatomist, that the human population includes individuals who have oval shaped heads and individuals who have round shaped heads (when viewed from above), did not form the basis of his conclusion that Offender One and the appellant each have oval shaped heads. That conclusion was based on Professor Henneberg's subjective impression of what he saw when he looked at the images. This observation applies to the evidence of each of the characteristics of which Professor Henneberg gave evidence.
>
> …
>
> Professor Henneberg's evidence gave the unwarranted appearance of science to the prosecution case that the appellant and Offender One share a number of physical characteristics. Among other things, the use of technical terms to describe those characteristics – Offender One and the appellant are both ectomorphic – was apt to suggest the existence of more telling similarity than

20 *Honeysett v R* [2013] NSWCCA 135 (5 June 2013) (Macfarlan JA, Campbell J and Barr AJ agreeing).

21 *Honeysett v The Queen* [2014] HCA 29 (13 August 2014), [43]-[46] (French CJ, Kiefel, Bell, Gageler and Keane JJ). The appellant's conviction was ordered to be quashed and a new trial allowed.

> to observe that each appeared to be skinny. Professor Henneberg's opinion was not based wholly or substantially on his specialised knowledge within s 79(1). It was an error of law to admit the evidence.

Facial mapping has been accepted as a form of biometric evidence, though with some reservations about the strength of expert opinions in particular cases. Body mapping has not been definitively accepted as scientifically reliable, and the few cases in which it has been considered in depth have cast doubt on the reasoning processes involved.

The courts' treatment of facial and body mapping as fields of 'specialised knowledge' has been criticised. In relation to the *Honeysett* case, Edmond and San Roque (2014, p. 324) have argued:

> We contend that too much weak, speculative and unreliable opinion is allowed into criminal proceedings, particularly in New South Wales. The problems with the contested image comparison evidence in *Honeysett* are representative of widespread problems with forensic science evidence more broadly. Following an extended review of the forensic sciences, involving submissions and hearings, a committee of the National Research Council of the United States National Academy of Sciences concluded that:
>
> With the exception of nuclear DNA analysis … no forensic method has been rigorously shown to have the capacity to consistently, and with a high degree of certainty, demonstrate a connection between evidence and a specific individual or source. … The simple reality is that the interpretation of forensic evidence is not always based on scientific studies to determine its validity. This is a serious problem.

Nonetheless, there is merit in challenging the scientific basis of new forensic techniques such as facial and body mapping, in order to ensure that the best evidence is presented before the courts. While this may not result in exclusion of expert opinion evidence, it may affect the weight it is given in the overall context of criminal proceedings.

Fingerprinting

As discussed in Chapter 2, fingerprinting has been routinely used by police in criminal investigations since the beginning of the 1900s (Coyle, Field & Wenderoth, 2009; Gans, 2011). Crime scene examiners may find 'latent' fingerprints or palm prints on objects, which can be visualised using laboratory processes. The prints can then be compared with those taken from a suspect or by searching for a match against a database of prints. This can be done in an automated way, for example, using the IDENT1 national fingerprint database that operates in the United Kingdom.

Courts around the world have routinely admitted fingerprint evidence in criminal proceedings for over a century.[22] Typically, the expert witness in such cases is an investigating police officer with specialised knowledge of fingerprinting techniques, or a forensic analyst, who was involved in the fingerprint collection and comparison process used in the investigation.[23]

Collection and comparison

The collection of fingerprints at a crime scene and their comparison to those taken from a suspect or found on a forensic database are regulated by forensic procedures legislation. The following criminal procedure legislation provides an example:[24]

> 3ZJ Taking fingerprints, recordings, samples of handwriting or photographs
>
> (1) In this section and in sections 3ZK and 3ZL:
>
> ***"identification material"***, in relation to a person, means prints of the person's hands, fingers, feet or toes, recordings of the person's voice, samples of the person's handwriting or photographs (including video recordings) of the person, but does not include tape recordings made for the purposes of section 23U or 23V.
>
> (2) A constable must not:
>
> (a) take identification material from a person who is in lawful custody in respect of an offence except in accordance with this section; or
> (b) require any other person to submit to the taking of identification material, but nothing in this paragraph prevents such a person consenting to the taking of identification material.
>
> (3) If a person is in lawful custody in respect of an offence, a constable who is of the rank of sergeant or higher or who is for the time being in charge of a police station may take identification material from the person, or cause identification material from the person to be taken, if:

22 In a 1912 case, it was observed: 'Signatures have been accepted as evidence of identity as long as they have been used. The fact of the individuality of the corrugations of the skin on the fingers of the human hand is now so generally recognised as to require very little, if any, evidence of it, although it seems to be still the practice to offer some expert evidence on the point. A finger print is therefore in reality an unforgeable signature': *Parker v R* [1912] HCA 29; (1912) 14 CLR 681, Griffith CJ at 683, cited in *R v Mitchell* [1997] ACTSC 93; (1997) 130 ACTR 48 (18 November 1997).

23 See, for example, the cases of *R v Regan* [2014] NSWDC 118 (16 June 2014); and *DPP v Watts* [2016] VCC 1726 (23 November 2016).

24 Part ID the *Crimes Act 1914* (Cth). Taking a fingerprint is classified as a 'non-intimate forensic procedure' which can be carried out with consent or by order of a senior police officer or magistrate on person in custody where other conditions are satisfied.

(a) the person consents in writing; or
(b) the constable believes on reasonable grounds that it is necessary to do so to:
   (i) establish who the person is; or
   (ii) identify the person as the person who committed the offence; or
   (iii) provide evidence of, or relating to, the offence; or

(ba) both of the following apply:

   (i) the identification material taken, or caused to be taken, is fingerprints or photographs (including video recordings) of the person;
   (ii) the offence is punishable by imprisonment for a period of 12 months or more; or
(c) the constable suspects on reasonable grounds that the person has committed another offence and the identification material is to be taken for the purpose of identifying the person as the person who committed the other offence or of providing evidence of, or relating to, the other offence.

(4) A constable may use such force as is necessary and reasonable in the circumstances to take identification material from a person under this section ....[25]

Police taking fingerprints under this provision may do so with or without the consent of the suspect. However, if there is a failure of compliance with the requirements of this section, or others that relate to the treatment of persons in custody and the taking of forensic samples, the defence is entitled to challenge the admissibility of the evidence based on the manner in which it was obtained. An example is a case involving the North Korean transport ship *Pong Su*, in which fingerprints of a suspect were taken by police officers. The defence argued that the circumstances in which the fingerprints were taken were oppressive in that the suspect 'had been exposed to the elements for two days prior to being taken into custody during which time he had no access to food and limited access to water and was found by police to be tired'. It was also submitted that the fingerprints had been illegally obtained due to non-compliance with the above provision (s3ZJ). The judge, however, found that the police officers had acted reasonably and in good faith, and that '[a]t the most any breach was a failure to comply with a procedural requirement' that did not require exclusion of the evidence.[26]

25 Subsections dealing with persons under the age of 18 years are not reproduced here. Note that more restrictive conditions may apply to minors: *R v SA, DD and ES* [2011] NSWCCA 60 (28 March 2011); *Hibble v B* [2012] TASSC 59 (20 September 2012). See also *Watkins v The State of Victoria & Ors* [2010] VSCA 138 (11 June 2010), which considered whether police had used excessive force in taking fingerprints from a suspect.

26 *Pong Su (No. 2)* [2004] VSC 492 (6 December 2004), per Kellam J at [31].

The process of comparing fingerprints may occur manually or by automated means via a database. The proposition that each individual's fingerprints are unique appears to be a basic assumption behind forensic uses of fingerprint matching, and has not been displaced by scientific advancement to date. Fingerprint comparison differs from other forms of biometrics, such as DNA matching, in that it does not rely on match probabilities. This means its presentation as evidence of identity is considerably simpler. However, there is still a degree of judgment required in making visual comparisons. Some critics point out that this inevitably introduces a capacity for error (Edmond, 2015).

The following extract provides an example of the use of fingerprint comparison adduced by the prosecution in a criminal case involving burglary:[27]

> The real strength of the Crown case lay in the fingerprint evidence. Ms Lam, a crime scene investigator, attended the scene at about 4.45 pm. She found a number of fingerprints, including some left on the television set, and both photographed them and took tape lifts from them. Mr Comber, a fingerprint expert, gave evidence that he had compared a fingerprint lifted from the television set with a fingerprint identified as that of the accused on the National Automated Fingerprint Identification System ('NAFIS'). He found that the two prints had both been made by the middle finger of the same left hand. There was no challenge to Mr Comber's methodology or as to the accuracy of this conclusion. I found him to be an impressive witness and accepted his evidence. It was not suggested that the fingerprint obtained from NAFIS had been incorrectly attributed to the accused and I was satisfied beyond reasonable doubt that the print had been left on the television set when touched by the accused.

The probative value of a fingerprint or palm print match must be assessed in the context of all other evidence in a criminal trial, and it will be of greatest significance if there is no apparently innocent explanation for how it came to be left at a crime scene.[28] This kind of evidence therefore operates as part of a circumstantial case against the defendant.

The following example of a fingerprint comparison report tendered during police testimony relates to a match of prints left during a burglary, and a young defendant identified as 'JP':[29]

27 *R v Millard* [2006] ACTSC 56 (6 June 2006), [15]. See also *R v Fitzgerald* [2005] SADC 118 (25 August 2005).

28 An unusual case where the defence sought to have fingerprint evidence excluded entirely was an appeal in which the defence alleged that police had forged the defendant's fingerprint on a cheque: *Mickelberg v The Queen* [2004] WASCA 145 (2 July 2004).

29 *JP v Director of Public Prosecutions (NSW)* [2015] NSWSC 1669 (11 November 2015), [7]. The police witness had prepared a 'Certificate of Expert Evidence' under s 177 of the UEL legislation, stating his qualifications as an examiner and presenting his conclusions.

> During the course of my daily duties, I carefully compared all the finger and palm impressions appearing in the photographs bearing Forensic Case Number 2819499 with the finger and palm impressions of [JP] born … as appearing on the fingerprint form by placing those photographs one at a time side by side with those finger and palm impressions and referring backwards and forwards between them. I compared pattern type and ridge flow, friction ridge characteristics, their relative positions to each other and the number of intervening ridges between those characteristics, that is the finger or palm prints appearing in the photographs bearing Forensic Case Number 2819499 against the finger or palm impressions of [JP] born … as appearing on the fingerprint form. The comparison process was carried out systematically and sequentially until all available friction ridge detail had been compared between the finger and palm impressions appearing in the photographs bearing Forensic case Number 2819499 and the finger and palm impressions of [JP] born … as appearing on the fingerprint form.
>
> Based wholly or substantially on my specialised knowledge and belief I am of the following opinion:
>
> - Graph W1 is identified to another person
> - Graph W2 is identified to another person
> - Graph W3 is identified to the Left Thumb of [JP] …
>
> That is to say the impressions appearing in the photographs bearing Forensic case Number 2819499 and labelled W3 are made by one of the same [JP] born …

Although match probabilities are not involved in fingerprint comparisons, the process of comparing two prints and arriving at a conclusion does involve the identification of numerous points of comparison, sometimes referred to as 'characteristic points'.[30] The more points that are compared, and the more similarity between the compared points, the more persuasive will be any conclusion drawn regarding identity. Although it is not strictly necessary for an expert witness to explicitly describe all of the details of the matching process in court testimony, cross-examination may be used by the defence to test the basis for concluding that fingerprints are the same.

The requirements for expert evidence mean that a witness with purported specialised knowledge should be able to explain how this provides a sound basis for

30 *JP v Director of Public Prosecutions (NSW)* [2015] NSWSC 1669 (11 November 2015), [36] referring to 'the case of *Bennett v Police* [2005] SASC 167 (4 May 2005) in which "more than 20 characteristics … were common and identical". In *JP*, the police witness claimed to have examined 35 comparison points but did not specify how many were considered to be a match with the defendant's prints, as opposed to the overall conclusion of identity.

the opinions arrived at in the case.[31] Where there are gaps in the explanations offered by the prosecution's experts, defence counsel may seek to have the opinion evidence excluded entirely, or ask for the jury to be cautioned in giving it weight as evidence.[32] It is also possible at the appeal stage for an appellant to argue that the fingerprint or other biometric evidence was not properly summarised by the judge in instructing the jury.[33]

It may also be possible to challenge inferences drawn from physical evidence, such as the estimated age of fingerprints. The time at which fingerprints were deposited through contact with an object may be of importance in assessing its relevance in a particular case. This will ordinarily involve additional forensic evidence.[34]

Another issue that judges must consider carefully is that a jury hearing that the defendant's fingerprints were matched to a crime scene using a police database may infer that the defendant has a criminal history, which explains the inclusion on the database. In such cases, the defence may seek to exclude evidence as unfairly prejudicial, or seek that the jury be discharged. A remedy is for the judge to warn the jury against making an adverse inference of this kind.[35]

DNA identification

Identification using DNA is generally regarded as more discriminating than any other biometric method. However, because it relies on the generation of a DNA profile from a biological sample, it can be susceptible to court challenges, for example, on the basis of sample integrity and the possibility of transference. Further complexities include the scientific processes and statistical interpretations involved (Gans & Urbas, 2002). Because DNA profiles are stored in increasingly large databases, new matching techniques allowing 'cold hits' and partial match searches are now used routinely (Smith & Mann, 2015). These issues will be discussed in turn, drawing on illustrative cases.

31 Leading authorities on specialized knowledge under UEL s79(1) are *Makita (Australia) Pty Ltd v Sprowles* [2001] NSWCA 305 (14 September 2001); *HG v The Queen* [1999] HCA 2; 197 CLR 414; and *Honeysett v The Queen* [2014] HCA 29; 253 CLR 122.

32 In *JP v Director of Public Prosecutions (NSW)* [2015] NSWSC 1669 (11 November 2015), defence arguments seeking exclusion of the fingerprint comparison evidence on the basis that the expert had insufficiently explained his reasoning process were unsuccessful. The judge noted at [43] that 'with fingerprint evidence it will often be the case that *"little explicit articulation or amplification"* of how the stated methodology warrants the conclusion that two fingerprints are identical will be required before it can be concluded that the second condition of admissibility under s 79(1) has been satisfied' (emphasis original), citing *Dasreef Pty Ltd v Hawchar* [2011] HCA 21; 243 CLR 588.

33 *Ghebrat v The Queen* [2011] VSCA 299 (12 October 2011).

34 See *R v SMR* [2002] NSWCCA 258 (1 July 2002).

35 See, for example, the defence submission in *R v Ahola (No. 6)* [2013] NSWSC 703 (14 May 2013), [3]: 'The submission is that the jury would inevitably infer from the [police officer's testimony] that the accused is a person with a criminal record whose fingerprints were held by the police, prior to them being taken from him with regard to this matter. It is that inference that forms the foundation for the application of the discharge of the whole jury'. The submission was unsuccessful in this case.

Collection and analysis

Compliance with forensic procedures legislation is a general pre-condition to the admissibility of DNA evidence. This applies to the taking of forensic material such as hair samples or buccal swabs from suspects, arrested persons and others.[36] Alternatively, the defence can challenge the integrity of samples collected and stored by police, on the basis that 'chain of custody' requirements have not been observed. This can support a hypothesis of 'contamination' (Edwards, 2006; Findlay & Grix, 2003).

As was discussed in Chapter 3, one of the most striking contamination cases worldwide is that of Farah Jama, who was wrongly convicted of rape on the basis of a DNA match. This was subsequently found to have been most likely a result of accidental contamination of the alleged rape evidence with a sample of Jama's DNA which was in the same forensic laboratory having been taken the day before during an unrelated investigation (Rayment, 2010; Cashman & Henning, 2012; Krone, 2012). Cases such as this reinforce the need for compliance with forensic collection, storage and analysis ('chain of custody') protocols, as errors can be very hard to identify and correct at trial.

In addition to accidental contamination, it is possible for DNA evidence to be manipulated by deliberate interference. In another case, defence lawyers suggested that the presence of an assault victim's blood on the clothing of the defendant may have been the result of either contamination or deliberate interference in a police facility, as one of its experts discerned 'post-transfusion' artefacts in a tested clothing sample, indicating that the blood involved may have come from the victim after police had taken a blood sample in the hospital rather than as a result of the alleged attack (Haesler, 2006).[37]

It should also be recalled (as discussed in Chapter 3) that a DNA match will ordinarily only have legal significance if there is no innocent explanation for the DNA being found where it was. Finding the defendant's DNA at the crime scene will normally be of little or no relevance if that happens to be the defendant's own home or workplace. However, even if it is a location where the defendant's DNA might not be expected to be found, there may be an innocent explanation for its presence. One explanation, often favoured by defence lawyers, is 'transference'.

36 See, for example, *Walker v Budgen* [2005] NSWSC 898 (7 September 2005); and *Hibble v B* [2012] TASSC 59 (20 September 2012) dealing with a DNA sample taken from a 13-year old suspect.

37 *R v Lisoff* [1999] NSWCCA 364 (22 November 1999). The court ruled that the matter was one that could be put before a jury for resolution, rather than require exclusion on the grounds of unfair prejudice to the defendant: "There is nothing so extraordinary about the conflict in the evidence presented in this case which would justify the conclusion that a careful and sensible jury, properly directed as to the relevant law and as to the relevant evidence, could not decide in a reasoned and responsible way whether or not the Crown had demonstrated beyond reasonable doubt that the body of evidence supporting the Crown case should be preferred to the opposed body of evidence" [64].

Because DNA is found in even small biological samples, it may be transferred through physical contact between persons or objects, and then onto other persons or objects. A person's DNA may be found at a location where he or she has never, or not recently, been. In order for the prosecution to be able to use the presence of DNA as proof of involvement in a crime, it may then be necessary to negate, beyond reasonable doubt, the possibility of transference. This situation has arisen in several noteworthy cases, including *Hiller*. The defendant in that case was charged with the murder of his estranged partner. Part of the prosecution's evidence was that his DNA was found on the deceased's pyjamas. However, none of the expert witnesses at the trial were able to rule out the possibility of transference, through the couple's children:[38]

> There is nothing in the evidence to exclude the possibility that the children may have had some of the appellant's DNA transferred to their sleeves or other parts of their clothing when they hugged him at the end of a week spent in his care, and then subsequently hugged their mother in a similar manner. Nor, is there any reason to suppose that DNA left on their clothing after contact with the appellant might not have been transferred to the deceased's pyjamas at some later stage when she had been handling that clothing.

This was the basis on which the murder conviction was quashed. However, on appeal by the prosecution it was overturned and a re-hearing of the appeal was ordered. This second appeal ordered a re-trial, at which the defendant elected to be tried by judge alone, rather than before a jury as in the first trial, and he was acquitted.[39]

In a more recent case, *Fitzgerald*, the transference problem was again raised by the defence in a murder trial. On the prosecution's case, the defendant's DNA was found on an object, a didgeridoo, in the house at which a fatal assault took place. Because the possibility of secondary transfer could not be ruled out, the court ultimately allowed an appeal and ordered that a verdict of acquittal be entered.[40]

In the 2013 case *Maryland v King*, the US Supreme Court upheld the use of DNA sampling in the criminal justice system against the Fourth Amendment of the US Constitution, which prohibits unreasonable searches and seizures, and requires warrants to be issued by a judge and supported by probable cause. King argued that the Maryland DNA Collection legislation violated the Fourth Amendment to the US Constitution. Although the Maryland Court of Appeals found the legislation was unconstitutional, the US Supreme Court held that taking DNA is a legitimate procedure to identify arrestees.

38 *Hillier v R* [2005] ACTCA 48 (15 December 2005), (Higgins CJ and Crispin P), [60].

39 The High Court appeal was *R v Hillier* [2007] HCA 13 (22 March 2007); which was followed by re-heard appeal in *Hillier v R* [2008] ACTCA 3 (6 March 2008); the final acquittal is unreported.

40 *Fitzgerald v The Queen* [2014] HCA 28 (13 August 2014).

The majority opinion considered that the Fourth Amendment permits police to undertake 'routine identification processes' in relation to arrestees,[41] including photographing and fingerprinting arrestees as part of the associated administrative process.[42] Further, that this is part of a legitimate 'need for law enforcement officers in a safe and accurate way to process and identify the persons and possessions they must take into custody'[43] and that DNA sampling is an extension of these more established methods.[44] Further, it considered that the cheek swap used to collect biological material was 'quick', 'painless' and 'no more invasive than fingerprinting'.[45]

According to the dissenting view in the case, the Supreme Court's finding promotes the collection of DNA by police from individuals that have not committed serious offences, or are even arrestees. Justice Scalia opined that the approach is a shift towards a 'genetic panopticon'[46] and '[n]o matter the degree of invasiveness, suspicionless searches are never allowed if their principal end is ordinary crime-solving'.[47]

Roth (2013, p. 298) argues that by running an arrestee's DNA profile against a database, seeking a 'cold hit' against DNA collected at the scenes of unsolved crimes, rather than a database of known offenders to establish his identity 'suggests that the state contemplates the arrest as a proxy for criminality rather than as a means of covering all those in custody whose identification needs confirmation'.

Scientific basis

The science underlying DNA identification has been extensively assessed in criminal proceedings around the world since the late 1990s. In a 2001 case that provides a representative example, the Profiler Plus DNA matching technology based on Polymerase Chain Reaction (PCR) analysis, that had been in use for over a decade, was found to be sufficiently accepted within the scientific community to be a valid means of identification in criminal trials. The judge stated:[48]

> The evidence in the present case was clear and, in my view, overwhelming. Whilst the Profiler Plus system is relatively new, it utilizes familiar technology for amplification and inspection of STR loci which technology is widely, almost universally, accepted in the relevant scientific community as reliable

41 *Maryland v. King*, 133 S. Ct. 1958, 1966 (2013), at 1976.
42 *Ibid,* quoting *Cnty. of Riverside v. McLaughlin*, 500 U.S. 44, 58 (1991).
43 *Ibid* at 1970.
44 *Ibid* at 1977.
45 *Ibid* at 1968.
46 *Ibid* at 1990.
47 *Ibid* at 1982.
48 *R v Karger* [2001] SASC 64 (29 March 2001), [229], [614] (Mullighan J) within a long and highly detailed judgment in which virtually every aspect of the technology was judicially considered. Although some of the primer sequences used in Profiler Plus had not been disclosed by the manufacturer, this was not regarded as an impediment to establishing reliability (Wiley & Hocking, 2003).

> and accurate. The variations fundamental to the Profiler Plus system, namely the particular loci and the number of them, the new primer sequences if they are new, and the use of Genotyper, have clearly been shown to have been accepted by the relevant scientific community as accurate and reliable. … The evidence overwhelmingly established that the Profiler Plus system is generally accepted throughout the forensic science community as reliable and accurate in DNA analysis for the purposes of human identification, including with low levels of DNA.

In many countries around the world, the main criteria for admissibility of opinion evidence from experts are those found in the 'specialised knowledge' provisions of evidence law rather than scientific criteria of reliability. This has developed from the US case *Daubert v Merrell Dow Pharmaceuticals, Inc.*[49] The *Daubert* standard is that there be a field of specialised knowledge, that the witness have such knowledge based on training, study or experience and that the opinions of the witness be wholly or substantially based on this. The forensic use of DNA in criminal investigations is now routinely accepted as a field of specialised knowledge (Gans & Urbas, 2002; Smith & Mann, 2015).

The first stage of DNA identification involves the generation of a profile from a crime scene and its comparison with the defendant's profile. The legal significance of the presence or absence of a match has been explained as follows:[50]

> A DNA profile taken from an evidence sample is compared to a sample provided by an individual. If the DNA profile taken from an evidence sample does not match the DNA profile of a person, then that individual can be conclusively excluded as being the source of the DNA from the evidence. If the profiles of the evidence sample and an individual do match, then there are two competing possibilities to explain the matching DNA profiles. The first possibility is that the DNA profile match has occurred because the DNA has originated from the person in question. The second possibility is that the DNA match has occurred by chance. That is, that there is someone else in the population who just happens to have the same DNA profile as the person in question. The probability of the evidence (i.e. probability of the sample matching the known or unknown person) given each of these scenarios is calculated using statistical analysis. A population database is used to provide an indication of the relevant prevalence of each of the alleles that were observed in the population. The given ratio of the two probabilities is called the likelihood ratio.

49 *Honeysett v The Queen* [2014] HCA 29 (13 August 2014). The Daubert case is the US Supreme Court decision of *Daubert v Merrell Dow Pharmaceuticals, Inc* [1993] USSC 99; 509 U.S. 579; 113 S.Ct. 2786; 125 L.Ed.2d 469; No. 92–102 (28 June 1993).

50 *Aytugrul v R* [2010] NSWCCA 272 (3 December 2010), [80] (McClellan CJ at CL) citing Sulan J in *R v Carroll* [2010] SASC 156 (28 May 2010), [28].

In other words, a DNA match only provides a link between the defendant and a crime on a probabilistic basis, whereas a non-match will exclude identification conclusively (Gans & Urbas, 2002). The significance of such a match will depend on the context and other evidence. However, this must be because any innocent explanation for the presence of the defendant's DNA at the crime scene, or on the body of a sexual assault victim, is excluded (Julian & Kelty, 2012; Julian et al., 2012). The analysis may be complicated further where the crime scene sample contains 'mixed profiles' indicating that it contains the DNA of more than one individual.[51]

Where there is considerable room for interpretation is in the significance and proper presentation of the statistical analysis that accompanies a DNA match (Goodman-Delahunty & Tait, 2006). This is because tools such as Profiler Plus only use a fixed number of markers or *loci* from the non-coding genetic sequences that are used in generating DNA profiles, meaning that two different individuals could have the same profile within these parameters. This then allows analysts to say that a match was found, such as between a crime scene sample and one taken from the defendant during investigation, and to state the approximate probability of this match being a result not of commonality of origin but of a random match. This is typically expressed as a random match probability relative to the general population or a subset of it, based on a representative sample.

The composition and size of the sample used may be significant in supporting the inferences to be drawn. In practice, population sample databases of only a few hundred are accepted by the courts as being sufficiently discriminating to allow valid statistical inferences to be drawn:[52]

> Databases have been built up by which the probability that the DNA of another person within the general population would match the DNA of the deceased at particular genetic markers may be estimated … It is accepted that the precision of the figures produced from any data base is dependent upon the size of the sample; the larger the sample, the greater the precision in the figures produced. The database for the RFLP results was compiled from the testing of 500 people who had donated blood at the Red Cross Blood Bank … The statistical validity of databases compiled from as low as 100 to 150 people is supported by a number of eminent scientists and scientific bodies.

In the *Pantoja* case, a question arose about the appropriateness of using a general database of profiles taken from a multicultural society where a majority of the

51 *Tuite v The Queen* [2015] VSCA 148 (12 June 2015); and *R v Xie (No. 18)* [2015] NSWSC 2129 (28 July 2015).

52 *R v Milat* (1996) 87 A Crim R 446; see also *R v To* [2002] NSWCCA 247 (26 June 2002).

adults were of white European ethnicity, when the defendant was a member of a distinctive group (identified as South American Quechua Indians).[53] The court held that this did not matter, as it was the racial characteristics of the (unknown) offender that were relevant to the appropriateness of the statistical database, rather than the ethnicity of the defendant. However, the statistical validity of the database still had to be established, which led to a successful appeal, a re-trial and a second appeal in which the conviction was finally affirmed.[54]

Juror comprehension

A frequently discussed question is whether juries are capable of understanding complex scientific information such as biometric identification technology, and if they are to be required to evaluate such evidence, the forms in which it should be presented so as to best facilitate comprehension (Goodman-Delahunty & Wakabayashi, 2012). A starting point is the view that complexity alone should not preclude scientific evidence from being heard by a jury:[55]

> Juries are frequently called upon to resolve conflicts between experts. They have done so from the inception of jury trials. Expert evidence does not, as a matter of law, fall into two categories: difficult and sophisticated expert evidence giving rise to conflicts which a jury may not and should not be allowed to resolve; and simple and unsophisticated expert evidence which they can. Nor is it the law, that simply because there is a conflict in respect of difficult and sophisticated expert evidence, even with respect to an important, indeed critical matter, its resolution should for that reason alone be regarded by an appellate court as having been beyond the capacity of the jury to resolve.

However, there are recognised dangers in the presentation of statistical identification evidence, such as the 'prosecutor's fallacy', which courts have had to consider (discussed in Chapter 3). This fallacy, so called because it tends to assist the prosecution rather than the defence, involves misstating the estimated frequency of the defendant's DNA profile (for example, 1 in 1 million) as the likelihood that the defendant left the crime scene DNA (which in a population of 23 million will be a very different from 1 million to 1). The case of *Keir* resulted in a quashed

53 *R v Pantoja* [1996] NSWSC 57 (1 April 1996).

54 *R v Pantoja* [1998] NSWSC 565 (5 November 1998).

55 *Velevski v The Queen* (2002) 76 ALJR 402; [2002] HCA 4, [182] (Callinan and Gummow JJ). This case did not concern DNA evidence but rather knife wounds and expert opinion as to how they could have been inflicted. Although expert opinion evidence is largely governed by s79 of the UEL, s80 does allow specialised knowledge to be supplemented by 'common knowledge' as part of the expert's reasoning, as it provides: 'Evidence of an opinion is not inadmissible only because it is about: (a) a fact in issue or an ultimate issue, or (b) a matter of common knowledge'. Thus, an expert witness may 'have regard to matters that are within the knowledge of ordinary persons in formulating his or her opinion' (Gaudron J, [82]).

conviction on appeal, on the basis that the judge had committed the prosecutor's fallacy in summing up the evidence to the jury.[56]

With regard to DNA match probabilities, it has also been argued that mathematically equivalent ways of expressing the same information can have different levels of persuasiveness to a jury. In the case of *Aytugrul*, the following evidence was at issue (Urbas, 2012):[57]

> A hair found on the deceased's thumbnail had been subjected to mitochondrial DNA testing. The results of that testing showed two things: first, that the appellant could have been the donor of the hair and, second, how common the DNA profile found in the hair was in the community. This second aspect of the results was expressed in evidence both as a frequency ratio and as an exclusion percentage. The expert who had conducted the test gave evidence to the effect that one in 1,600 people in the general population (which is to say the whole world) would be expected to share the DNA profile that was found in the hair (a frequency ratio) and that 99.9 per cent of people would not be expected to have a DNA profile matching that of the hair (an exclusion percentage).

The defence sought to argue that there was unfair prejudice in putting the percentage before the jury, as this was overly persuasive and invited a subconscious 'rounding up' to 100 per cent certainty. However, this was not accepted by the relevant court given the expert's explanations:[58]

> The unfair prejudice said to arise in this case was alleged to flow from the use of a percentage figure, which carried a "residual risk of unfairness deriving from the subliminal impact of the raw percentage figures" by way of rounding up the percentage figure to 100. If the exclusion percentage were to be examined in isolation, the appellant's arguments appear to take on some force. But to carry out the relevant inquiry in that way would be erroneous. In this case, both the frequency ratio and the manner in which the exclusion percentage had been derived from the frequency ratio were to be explained in evidence to the jury. The risk of unfair prejudice – described by the appellant as the jury giving the exclusion percentage "more weight … than it deserved" – was all but eliminated by the explanation.

56 *R v Keir* [2002] NSWCCA 30 (28 February 2002). The defendant was convicted on the re-trial, and an appeal against that conviction was unsuccessful: *Keir v R* [2007] NSWCCA 149 (6 June 2007). The prosecutor's fallacy was also discussed in *Aytugrul v R* [2010] NSWCCA 272 (3 December 2010).

57 *Aytugrul v The Queen* [2012] HCA 15 (18 April 2012), (French CJ, Hayne, Crennan and Bell JJ), [2] (note omitted after the words 'frequency ratio', as follows: 'Sometimes called a "random occurrence ratio" or a "frequency estimate"'). Heydon J agreed with the majority in a separate judgment.

58 *Aytugrul v The Queen* [2012] HCA 15 (18 April 2012), (French CJ, Hayne, Crennan and Bell JJ) [30] (note omitted).

The concurring judgment in *Aytugrul* suggested that the jury could be trusted to work out the statistical issues, even though they were difficult:[59]

> No doubt both the "frequency estimate" and the "exclusion percentage" evidence, like many other aspects of the expert evidence, were difficult for the jury to deal with. The field is arcane. But any criminal jury of 12 is likely to contain at least one juror capable of realising, and demonstrating to the other jurors, that the frequency estimate was the same as the exclusion percentage. Further, detailed evidence was given about how the "exclusion percentage" evidence was derived from the concededly admissible "frequency estimate" evidence, and how their significance was identical.

In a 2015 case, the question arose whether the scientific reliability of a new statistical technique applied to the analysis of small amounts of DNA used in matching was a matter going to the admissibility of expert evidence. The court held that it does not, but rather affects the probative value of the evidence and the potential prejudicial effect that its presentation to the jury may have. In this case, the appeal judges agreed with the conclusions reached by the trial judge in relation to both aspects of the evidence, holding that the probative value of the evidence was not outweighed by the alleged prejudicial effect:[60]

> In my view, the DNA evidence viewed as a whole is highly probative. It may be used by a jury to put the accused both inside and outside the house on the night in question. This is so, notwithstanding only small amounts of DNA matching that of the accused were found on the relevant items inside the house, and that other people also contributed to the DNA found on these items. The limitations in the STRmix methodology acknowledged by the prosecution witnesses must have some effect on the quality of the DNA evidence. However, I am not persuaded that they erode its probative value to any significant degree. Whilst the amounts of DNA may be small in some cases, the fact that DNA matching the accused's was found on a number of items both inside and outside the house in my view fortifies the overall probative value of the DNA evidence, which I assess to be high.
>
> In this case, the danger of unfair prejudice is said to arise from a particular issue identified by Ms Taupin in the STRmix analysis of Item 1–2, although it has wider consequences as it is a product of the way in which STRmix works generally. Ms Taupin identified, having closely examined the STRmix case-notes, that at two of the 10 markers the probability of the evidence given the prosecution hypothesis was very low, yet the likelihood ratios for the markers favoured the prosecution hypothesis. Ms Taupin pointed out that this means that STRmix produces likelihood ratios strongly favouring the prosecution

59 *Aytugrul v The Queen* [2012] HCA 15 (18 April 2012), [75] (Heydon J).
60 *Tuite v The Queen* [2015] VSCA 148 (12 June 2015), [122]-[124].

> hypothesis in circumstances where there is only very weak evidence to support that hypothesis. That, in combination with the very high likelihood ratios generated by STRmix, is said to be unfairly prejudicial to the accused and not something that should be allowed in a criminal trial.

A more subtle problem relating to juror comprehension is sometimes referred to as the 'CSI effect' by reference to a popular television series depicting forensic science in investigations (Wise, 2010). The perceived problem is that, even where experts provide accurate information about the limitations and confidence levels of their analysis, the jury may still be overwhelmed by the scientific nature of the evidence and give it more weight than it deserves.[61] The same may be true where evidence such as a DNA match is of marginal probative value in a prosecution:[62]

> Moreover, one of the dangers associated with DNA evidence, is what has come to be known as the 'CSI effect'. The 'CSI effect' is a reference to the atmosphere of scientific confidence evoked in the imagination of the average juror by descriptions of DNA findings. As we have explained, as a matter of pure logic, the DNA evidence has little or no probative value. By virtue of its scientific pedigree, however, a jury will likely regard it as being cloaked in an unwarranted mantle of legitimacy – no matter the directions of a trial judge – and give it weight that it simply does not deserve. The danger of unfair prejudice is thus marked, and any legitimate probative value is, at best, small.

Conversely, jurors exposed to fictional representations of forensic science may unrealistically expect to be presented with DNA or other biometrics in every case, and when this does not eventuate, may wrongly view this as a defect in the prosecution's case (Roux et al., 2012; Whiley & Hocking, 2003; Meyers, 2007).[63]

DNA databases

As increasing numbers of DNA profiles have been collected during criminal investigations, these have been stored on police databases, leading to the possibility of repeated use including by searching for a 'cold hit' between a crime scene sample and a profile already added to the database.[64] Forensic databases have

61 The 'CSI effect' has also been referred to as the 'white coat' effect: *Morgan v R* [2011] NSWCCA 257 (1 December 2011), [145], cited in *R v MK* [2012] NSWCCA 110 (4 June 2012).

62 *DPP v Wise (a pseudonym)* [2016] VSCA 173 (21 July 2016), [70]; *DPP v Massey (a pseudonym)* [2017] VSCA 38 (6 March 2017), (Weinberg JA), [24].

63 *R v Drummond (No. 2)* [2015] SASCFC 82 (5 June 2015).

64 See, for example, *Sleiman v Murray* [2009] ACTSC 82 (15 July 2009); and *R v Smith [No 1]* [2011] NSWSC 725 (26 May 2011).

become a powerful investigative tool (Smith, 2016). Not surprisingly, then, the conditions under which a defendant's profile may have been obtained, stored and retained on a DNA database have become the subject of scrutiny in criminal trials.

Forensic procedures legislation governs how forensic samples stored in DNA databases may be used. The legislation distinguishes between volunteers, arrested persons and convicted persons, with different requirements applying to each class. Permissible matching is regulated by matching tables in the legislation. Finally, requirements relating to use and removal of profiles from forensic databases are set out in detail. The failure of police or other officials to comply with the requirements of forensic procedures legislation can readily lead to exclusion of evidence obtained from a stored DNA profile.[65]

The unlawful retention of DNA profiles on databases was at issue in the landmark *Marper* case in the United Kingdom.[66] Two individuals, one of them a 12-year-old, whose profiles had been entered on the database when they were arrested for a reportable offence, sought to have them removed when they were not convicted. The House of Lords found in favour of the police, arguing that the retention was lawful under applicable legislation, but the European Court of Human Rights ruled otherwise, holding that the 'blanket and indiscriminate nature' of the retention regime under the legislation did not strike a proper balance between public and private interests (Smith, 2016).

A further consideration regarding the use of forensic DNA databases is the possibility of searching for partial matches, also known as 'familial searching'. This involves recording and investigating matches that nearly but do not fully coincide, and so cannot be from the same individual. However, it may be that the crime scene sample came from a close relative of someone who is on the DNA database, which provides an investigative lead even when the actual offender's profile is not on the database. This significantly extends the scope of 'cold hit' matching processes, and has been used to solve serious crimes in other countries. In many jurisdictions, forensic procedures legislation does not specifically regulate partial matching, but appears to allow its use (Smith & Urbas, 2012).

References

Buckland, P. (2014). *Honeysett v The Queen* (2014): Opinion evidence and reliability: A sticking point. *Adelaide Law Review 35*(2), 449.

Cashman, K. & Henning, T. (2012). Lawyers and DNA: Issues in understanding and challenging the evidence. *Current Issues in Criminal Justice 24*, 69.

65 Examples include *Hibble v B* [2012] TASSC 59 (20 September 2012); *R v Dean* [2006] SADC 54 (25 May 2006).

66 *R v Marper and S* (2002) EWCA Civ 1275 (Court of Appeal); *R v Marper and S* (2004) UKHL 39 (House of Lords); *Case of S and Marper and the United Kingdom* [2008] ECHR 1581.

Coyle, I., Field, D. & Wenderoth, P. (2009). Pattern recognition and forensic identification: The presumption of scientific accuracy and other falsehoods. *Criminal Law Journal 33*, 214.

Edmond, G. (2015). What lawyers should know about the forensic "sciences". *Adelaide Law Review 37*, 33.

Edmond, G. & San Roque, M. (2012). The cool crucible: Forensic science and the frailty of the criminal trial. *Current Issues in Criminal Justice* 24(1), 51.

Edmond, G. & San Roque, M. (2014). Honeysett v The Queen: Forensic science, 'specialised knowledge' and the uniform evidence law. *Sydney Law Review 36*(2), 323.

Edwards, K. (2006). Cold hit complacency: The dangers of DNA databases re-examined. *Current Issues in Criminal Justice 18*(1), 92.

Findlay, M. & Grix, J. (2003). Challenging forensic evidence? Observations on the use of DNA in certain criminal trials. *Current Issues in Criminal Justice 14*(3), 269.

Gans, J. (2005). DNA identification and rape victims. *University of New South Wales Law Journal 28*(1), 272.

Gans, J. (2007a). Much repented: Consent to DNA sampling. *University of New South Wales Law Journal 30*(3), 579.

Gans, J. (2007b). Catching Bradley Murdoch: Tweezers, pitchforks and the limits of DNA sampling. *Current Issues in Criminal Justice 19*, 34.

Gans, J. (2007c). The Peter Falconio investigation: Needles, hay and DNA. *Current Issues in Criminal Justice 18*(3), 415.

Gans, J. (2011). A tale of two High Court forensic cases. *Sydney Law Review 33*(3), 515.

Gans, J. & Urbas, G. (2002). DNA evidence in the criminal justice system. *Trends and Issues in Crime and Criminal Justice No. 226*. 3. Canberra: Australian Institute of Criminology.

Goodman-Delahunty, J. & Tait, D. (2006). DNA and the changing face of justice. *Australian Journal of Forensic Sciences 38*, 97.

Goodman-Delahunty, J. & Wakabayashi, K. (2012). Adversarial forensic science experts: An empirical study of jury deliberation. *Current Issues in Criminal Justice 24*(1), 85.

Haesler, A. (2006). DNA in court. *Journal of the Judicial Commission of New South Wales. 8*(1), 121.

Julian, R. & Kelty, S. (2012). Forensic science and justice: From crime scene to court and beyond. *Current Issues in Criminal Justice 24*, 1.

Julian, R., Kelty, S. & Robertson, J. (2012). Get it right the first time: Critical issues at the crime scene. *Current Issues in Criminal Justice 24*(1), 25.

Krone, T. (2012). Raising the alarm? Role definition for prosecutors in criminal cases. *Australian Journal of Forensic Sciences 44*(1), 15.

Meyers, L. (2007). The problem with DNA. *Monitor on Psychology 38*, 52.

Rayment, K. (2010). Faith in DNA: The Vincent Report. *Journal of Law, Information and Science 20*(1), 238.

Roth, A. (2013). Maryland v. King and the wonderful, horrible DNA revolution in law enforcement. *Ohio State Journal of Criminal Law 11*, 295.

Roux, C., Crispino, F. & Ribaux, O. (2012). From forensics to forensic science. *Current Issues in Criminal Justice 24*(1), 7.

Smith, M. (2016). *DNA Evidence in the Australian Legal System*. Chatswood, NSW: LexisNexis Butterworths.

Smith, M. & Mann, M. (2015). Recent developments in DNA evidence. *Trends and Issues in Crime and Criminal Justice No. 506*. 1. Canberra: Australian Institute of Criminology.

Smith, M. & Urbas, G. (2012). Regulating new forms of forensic DNA profiling under Australian legislation: Familial matching and DNA phenotyping. *Australian Journal of Forensic Sciences 44*, 63.

Urbas, G. (2012). The High Court and the admissibility of DNA evidence: Aytugrul v The Queen [2012] HCA 15. *Canberra Law Review 11*(1), 89.

Whiley, D. & Hocking, B. (2003). DNA: Crime, law and public policy. *University of Notre Dame Australia Law Review 5*, 37.

Wise, J. (2010). Providing the CSI treatment: Criminal justice practitioners and the CSI effect. *Current Issues in Criminal Justice 21*(3), 383.

7

BIOMETRICS IN CRIMINAL APPEALS AND POST-CONVICTION REVIEWS

Introduction

This chapter discusses the ways in which biometric identification has featured in criminal appeals and other reviews of criminal convictions. Appellate review allows the criminal justice system to recognise and correct errors, including wrongful convictions and other miscarriages of justice. However, as appeal rights are limited, other forms of review also play an important role. Discussed in this chapter are innocence projects, judicial inquiries and review commissions that have used biometrics in their efforts to uncover the truth about past crimes.

Criminal appeals

The first avenue of redress for most convicted offenders who claim that they have been the subject of a miscarriage of justice is to lodge an appeal. Depending on conditions imposed on such applications, including time limits and leave requirements, this may result in a conviction being overturned. In general, appellate courts can then either order a re-trial, or order a different verdict. In those rare cases where actual innocence can be established, the only appropriate outcome is quashing of the conviction and its replacement with a verdict of acquittal.[1]

Criminal appeals took on a distinctive form in the early twentieth century with the establishment in the United Kingdom of the Court of Criminal Appeal in 1907 (Corns & Urbas, 2008). Many jurisdictions empower a court of appeal to overturn

1 Actual innocence need not be established in order for an appeal to succeed, nor is this often possible. Rather, it is sufficient that enough doubt is cast on the conviction that it must be regarded as 'unsafe and unsatisfactory': *M v R* (1994) 181 CLR 487; see also *Gipp v R* (1998) 194 CLR 106 and *Chidiac v R* (1991) 171 CLR 432.

a conviction according to the following 'common form' grounds (Urbas, 2002; Corns & Urbas, 2008):[2]

i that the verdict was unreasonable or unsupportable having regard to the evidence;
ii that there was an error of law; or
iii that on any other ground there was a miscarriage of justice.

Although biometric evidence can be involved in any of these grounds of appeal, the possibility of using new evidence to cast doubt on a criminal conviction is best supported under the third limb. Evidence will exculpate the appellant if it tends to show that someone else committed the crime.[3] The courts of appeal have the power to receive new evidence in an appeal against conviction, including appointing a person with special expert knowledge as an assessor.[4] If the evidence on appeal differs from that admitted at the trial, the appellate judges must make an independent assessment of the case against the appellant based on the new evidence.

Criminal appeals and biometrics

The most obvious way in which biometrics can feature in a criminal appeal is by way of linking someone other than the appellant to the crime. For example, where a conviction was based largely on eye witness identification rather than forensic analysis, new evidence such as DNA testing of samples retained from the investigation may yield powerful exculpatory evidence (ALRC, 2003: [45.1]). Even where forensic analysis was involved at the trial phase, later testing using improved techniques may yield different results by the time of a later appeal.[5] A court of appeal may remedy a miscarriage of justice, but requirements of obtaining leave and time limits may make this a difficult option for convicted persons to pursue. Additionally, the fact that only one appeal to a court of appeal is usually possible leaves unsuccessful appellants no other option but an appeal to the Supreme Court of the United States, the Supreme Court of the United Kingdom or the High Court of Australia. However, the High Court of Australia, for example, has consistently ruled that it is not a Court of Criminal Appeal and has no power to

2 Subject to the 'proviso' that the conviction may be allowed to stand if the court is of the opinion that notwithstanding that the appellant has made out one or more of these grounds, no substantial miscarriage of justice has occurred (Penhallurick, 2003): see, for example, *Criminal Appeal Act 1912* (NSW), s6(1).

3 *Button v The Queen* [2002] WASCA 35 (25 February 2002), discussed in Goldingham (2002).

4 See for example, *Criminal Appeal Act 1912* (NSW), s12. Appeals against sentence are not discussed here, but note that questions of finality and double jeopardy also arise in relation to re-sentencing (Urbas, 2012).

5 An example is the Queensland case involving Frank Button discussed later in this chapter.

receive new evidence, including DNA evidence (Hamer, 2015; Milne, 2015; see also Urbas, 2002).[6]

Second and subsequent appeals

In view of the limits on criminal appeals, some jurisdictions have engaged in law reform to allow second or subsequent appeals to its Court of Criminal Appeal (Sangha, 2015). Such a provision allows a higher court to hear an appeal against conviction even where there has already been a previous appeal, if satisfied that there is fresh and compelling evidence that should, in the interests of justice, be considered. These requirements are defined as follows:[7]

> Evidence relating to an offence is –
>
> a **"fresh"** if –
> b it was not adduced at the trial of the offence; and
> c it could not, even with the exercise of reasonable diligence, have been adduced at the trial; and
> d **"compelling"** if –
> e it is reliable; and
> f it is substantial; and
> g it is highly probative in the context of the issues in dispute at the trial of the offence.

The form that such evidence might take includes fresh and compelling biometric analysis. For example, a crime scene sample collected before the trial may not have been tested, or testing may not have yielded results, due to the limitations of forensic analysis at the time. With advances in techniques, such as testing using small or degraded biological samples (as discussed in Chapter 3), testing may become possible years afterwards. This could show that the convicted person is not the offender. By this time, the convicted person may have already appealed unsuccessfully. The new legislation allows a second or subsequent appeal using the exonerating biometric evidence (Sangha & Moles, 2015). This is the model used by some innocence projects, discussed later in this chapter.

Double jeopardy and appeals against acquittal

The criminal law has for centuries restricted the ability of the prosecution to appeal against acquittals, based on the precept that it is unjust to expose a person to

6 *Mickelberg v The Queen* (1989) 167 CLR 259; *Eastman v The Queen* (2000) 203 CLR 1; *Re Sinanovic's Application* [2001] HCA 40; (2001) 180 ALR 448.

7 *Criminal Law Consolidation Act 1935* (SA), s353A inserted by the *Statutes Amendment (Appeals) Act 2013* (SA); and *Criminal Code Amendment (Second or Subsequent Appeal for Fresh and Compelling Evidence) Act 2015* (Tas).

punishment more than once in relation to the same crime. This is encapsulated in the rule against double jeopardy (MCCOC, 2003; Burton, 2004; Cowdery, 2005; Griffith & Roth, 2006) operating through the pleas of *autrefois convict* and *autrefois acquit*.[8]

Historically, the impetus for reform of double jeopardy laws has arisen from specific high profile cases. In Australia, for example, these arose largely in response to a child murder case in Queensland. Convicted of the murder of a 17-month-old baby in 1985, partly on the basis that his distinctive teeth were matched to a bite mark on the body of the victim, Raymond Carroll appealed successfully, so that the Queensland Court of Appeal quashed the conviction and entered a verdict of acquittal. This meant that a second prosecution for murder was precluded by double jeopardy rules. However, Carroll had given evidence at his trial denying involvement in the abduction and killing of the child, and on the basis of improved forensic odontological methods, the prosecution brought a charge of perjury. He was convicted on that second charge in 2000, on a jury verdict, and again appealed successfully, with the Court of Appeal accepting that the perjury conviction was in essence a re-trial of the murder case under a different charge. The High Court agreed, meaning that Carroll could never be re-convicted.[9] Public dissatisfaction with this outcome together with some academic and political support for a change in the law led to the enactment of legislation allowing appeals against acquittals in limited circumstances (Corns, 2003; Burton, 2004).

In Queensland, the provision applies only to a re-trial for murder where there is fresh and compelling evidence against the acquitted person and it is in the interests of justice to overturn the acquittal and order a re-trial.[10] In New South Wales, an application may be made in relation to any life sentence offence, including murder and certain drugs and sexual offences. However, there have been no murder re-convictions following an overturned acquittal to date.[11]

Several jurisdictions have adopted similar double jeopardy reforms, preceded by changes to double jeopardy laws in the United Kingdom (MCCOC, 2003), allowing re-trials after Crown appeals against acquittal.[12] The basis for these

8 See, for example, *Criminal Procedure Act 1986* (NSW), s156.

9 *The Queen v Carroll* (2002) 213 CLR 635. This was a prosecution appeal following the Court of Appeal decision.

10 *Criminal Code*, Chapter 68, added by the *Criminal Code (Double Jeopardy) Amendment Act 2007* (Qld).

11 *Crimes (Appeal and Review) Act 2001 (NSW), Part 8, added by the Crimes (Appeal and Review) Amendment (Double Jeopardy) Act 2006* (NSW). These provisions have been considered in *R v PL* [2009] NSWCCA 256 (8 October 2009); *Atkins v Attorney General of New South Wales* [2016] NSWSC 1412 (12 October 2016). There has also been public disquiet about the so-called 'Bowraville murders' case, with political pressure to use the double jeopardy reforms in NSW to re-open the acquittal of a key suspect, based on a novel argument that evidence is to be considered fresh due to a change in its admissibility after amendments to the *Evidence Act 1995* (NSW): http://www.ruleoflaw.org.au/double-jeopardy-bowraville-murders

12 By contrast, in the United States the rule against double jeopardy is a constitutional safeguard that cannot be abrogated by federal or state legislation (Thomas, 1998; Rudstein, 2004).

reforms was explained as follows by Lord Justice Auld who conducted a review of that country's court system (Auld, 2001) and posed the following questions:

> If there is compelling evidence ... that an acquitted person is after all guilty of a serious offence, then, subject to stringent safeguards ..., what basis in logic or justice can there be for preventing proof of that criminality? And what of the public confidence in a system that allows it to happen?

Similar to the laws allowing second and subsequent appeals against convictions in some jurisdictions, appeals against acquittal are generally limited to those cases in which there is fresh and compelling evidence of guilt, which could be in the form of new or improved biometric identification. For example, the evidence in an initial prosecution case may be insufficient to identify the accused as the offender beyond reasonable doubt. Later biometric analysis might provide a more conclusive link, which together with the other available evidence, might then be sufficient to safely convict the accused.[13]

However, post-conviction or post-acquittal testing depends on the preservation of evidence that can be tested (Urbas, 2002; Weathered, 2003; Weathered & Blewer, 2009; Hamer, 2014). As noted later in relation to the *Chamberlain* case, the destruction of forensic samples during or after laboratory testing can deny access to post-trial testing. This has led to calls for legislative requirements for sample preservation (ALRC, 2003). Despite the possibility of appeals based on fresh and compelling evidence, either against a conviction or an acquittal, these legal mechanisms appear to be rarely exercised in practice (Hamer, 2014).

Post-conviction reviews

Innocence projects

The potential for remedying wrongful convictions with the help of biometrics such as DNA identification has been the impetus for the establishment of many innocence projects, which are usually based in universities, as discussed in Chapter 3 (Hamer, 2014):[14]

> Fortunately, DNA profiling technology can provide strong proof of factual innocence. If biological material believed to be that of the perpetrator is available, and a DNA profile from that material does not match the DNA profile of the defendant, this provides practical certainty that the defendant is not the perpetrator. The strength of DNA profiling evidence in such cases is

13 Though note that there is some doubt about whether DNA evidence on its own could ever be sufficient for a conviction: see Ligertwood (2011) and the case of *Forbes v The Queen* [2010] HCATrans 120 (18 May 2010).

14 Notes omitted. See also Christian (2001); De Foore (2002); Urbas (2002) and Weathered (2004).

> quite exceptional. Generally it is just as difficult achieving certainty about innocence as it is about guilt. For this reason, innocence projects generally limit themselves to cases where DNA may be available.

The original innocence project was established in the United States. Based in the Cardozo Law School in New York, it has made over 350 exonerations using DNA evidence. In addition, the work of this and similar bodies has been instrumental in identifying and addressing the causes of wrongful convictions, with inaccurate eyewitness identification (discussed in Chapter 6) being the leading cause:[15]

> Eyewitness misidentification is the greatest contributing factor to wrongful convictions proven by DNA testing, playing a role in more than 70 per cent of convictions overturned through DNA testing throughout the United States.

The University of Bristol has been the leading innocence project in the United Kingdom and operated a specialist pro bono clinic from 2005 to 2015. Since that time there have been over 30 other innocence projects established at universities throughout England and Wales.[16] In Australia, similar bodies have been set up at Griffith University, Edith Cowan University and the University of Technology in Sydney.[17] Most of these follow the emphasis on post-conviction DNA testing shown to have been successful in overseas jurisdictions such as the United States.

The model of the innocence project has been followed in some cases by the establishment of an administrative body by governments to review claimed miscarriages of justice. The following provides an example of the functions of one such body, set out in legislation:[18]

a to consider any application under this Division from an eligible convicted person and to assess whether the person's claim of innocence will be affected by DNA information obtained from biological material specified in the application,
b to arrange, if appropriate, searches for that biological material and the DNA testing of that biological material,

15 See, for example, https://www.innocenceproject.org/causes/eyewitness-misidentification Other causes of wrongful conviction include false confessions, investigation or prosecution misconduct, poor defence representation, and forensic errors.
16 University of Bristol Law School. Retrieved from http://www.bristol.ac.uk/law/study/law-activities/innocenceproject
17 See, for example, https://www.griffith.edu.au/criminology-law/innocence-project; http://www.ecu.edu.au/schools/arts-and-humanities/research-and-creative-activity/sellenger-centre-for-research-in-law-justice-and-social-change/criminal-justice-review-project/overview; https://www.uts.edu.au/research-and-teaching/our-research/law-research-centre/about-us/history
18 *Crimes (Appeal and Review) Amendment (DNA Review Panel) Act 2006* (NSW), since repealed, added provisions establishing the panel to the *Crimes (Appeal and Review) Act 2001* (NSW). These were then removed by the *Crimes (Appeal and Review) Amendment (DNA Review Panel) Act 2013* (NSW) with effect from 23 February 2014.

c to refer, if appropriate, a case to the Court of Criminal Appeal under this Division for review of a conviction following the receipt of DNA test results, and

d to make reports and recommendations to the Minister on systems, policies and strategies for using DNA technology to assist in the assessment of claims of innocence (including an annual report of its work and activities, and of statistical information relating to the applications it received).

Persons meeting the description of 'eligible convicted offender' were provided with the opportunity to apply to the review panel to have their cases considered. This term was defined to include those serving sentences of 20 years or more, whether still in custody or on parole, or those whose 'special circumstances' warranted the application. Importantly, the application had to make a case that DNA information would assist in exonerating the person:[19]

> A convicted person is eligible to make an application to the Panel if, and only if, the person's claim of innocence may be affected by DNA information obtained from biological material specified in the application.

This application was to be assessed by the six member panel, which included a former judicial officer, a representative of the Attorney-General's Department, a victims' representative, a police representative and prosecution and defence lawyers. This review panel ceased operations in 2014, apparently not having referred any cases to a court of criminal appeal for review (Hamer & Edmond, 2013).

Judicial inquiries and commissions

Historically, the task of correcting miscarriages of justice fell to the executive rather than the judiciary, at least until the creation of the Court of Criminal Appeal (Spencer, 1982). The main mechanism used was the pardon. The prerogative of mercy is generally preserved under statute, which often also contains provisions allowing the establishment of reviews such as judicial enquiries into suspected miscarriages of justice (Caruso & Crawford, 2014). The long-standing institution of the Royal Commission can also be used to investigate alleged wrongful convictions. The result of an inquiry may be the pardon and release of the imprisoned person.

Although these are powerful mechanisms for the correction of miscarriages of justice, they are established on an *ad hoc* basis, often only after years of public agitation, and there thus is no predictability that such a body will be available in every case. This has led some observers to call for a Criminal Cases Review Commission (CCRC), based on models developed in the United Kingdom, as discussed in

19 *Crimes (Appeal and Review) Amendment (DNA Review Panel) Act 2006* (NSW), inserting s89 (since repealed) into the *Crimes (Local Courts Appeal and Review) Act 2001* (NSW).

Chapter 3 (Weathered & Blewer, 2009; Hamer, 2014). This has been discussed as follows (Weathered, 2013, p. 450):

> The most comprehensive body created to correct wrongful convictions is the Criminal Cases Review Commission ('CCRC') based in Birmingham, UK, which operates for England, Wales and Northern Ireland (for relevant legislation in England, Wales and Northern Ireland, see the *Criminal Appeal Act 1995* (UK) c 35, s 8; see also the Ministry of Justice, Criminal Cases Review Commission website at <http://www.ccrc.gov.uk>).
>
> Scotland and Norway have also each established their own CCRC, while other countries including Australia are still considering whether to create such a body. The CCRC is an independent, government-funded body that investigates claims of miscarriages of justice with the ability to refer cases to their courts of appeal. DNA innocence testing is incorporated within its broad and extensive powers of investigative review.

The establishment of a CCRC in the United States or Australia would need to navigate the federal system of criminal laws and courts, so that cases arising in each state would be referred to that particular jurisdiction's relevant appellate court.

Miscarriage of justice cases

The remainder of this chapter reviews some of the most significant miscarriage of justice cases in Australia where forensic evidence, such as biometrics, played a substantial role either in the initial prosecution, or in the appeal or other review that followed it. The expression 'miscarriage of justice' is used to refer to 'a false attribution of guilt, that is, finding someone guilty who was actually innocent' (Young, 2010). In serious cases, this leads to wrongful imprisonment (Zdenkowski, 1993). Miscarriage of justice is to be distinguished from a conviction that is overturned because of some procedural error at trial, such as a wrong decision on a question of admissibility of evidence (Spencer, 1992). The cases discussed below are the relatively few exonerations in Australia based on extensive review, either by a court or another review mechanism.[20]

Colin Campbell Ross

In 1922, Colin Campbell Ross was convicted of the murder of 12-year-old Alma Tirtschke. After a jury trial, he was convicted and sentenced to death. He then appealed unsuccessfully to higher courts. The case was as follows:[21]

20 Further literature on miscarriages of justice in the United Kingdom and the United States is discussed by Roach (2015).

21 *Ross v The King* (1922) 30 CLR 246 (Knox C.J., Gavan Duffy and Starke JJ, with Higgins J concurring). Isaacs J delivered a dissenting judgment.

> In the present case, the nude body of a young girl, twelve years of age, was found lying dead in an alley off Little Collins Street, Melbourne. Medical evidence disclosed that the cause of death was strangulation from throttling, that there were bruises and abrasions which indicated violence, and that there was a recent tear at the lower border of the hymen which passed completely through the hymen into the tissue of the vaginal wall. Evidence was also adduced by the Crown from which a jury might infer that this child had gone into an arcade known as the Eastern Arcade, in which the prisoner had a wine saloon, that she was there enticed by the prisoner into his wine saloon and was carnally known and killed by him. The prisoner, who gave evidence on his own behalf, did not suggest that he had killed the child in circumstances that might reduce the act from one of murder to one of manslaughter. He admitted that he had noticed a young girl, similar in appearance to the dead child, in the Arcade; but he denied that he had spoken to her or that she had been in his wine saloon, and he denied that he had anything to do directly or indirectly with the death of the murdered child. The jury found the prisoner guilty of the murder of the child.

On appeal, reference was made to 'evidence which went to identify the hair of the dead child with that found on certain blankets', but this was not pivotal in the Court's decision. Rather, the majority accepted that the trial judge had given correct directions to the jury on issues including an alleged confession by the accused. Special leave to appeal was therefore rejected by the High Court, and the sentence of execution was carried out a few weeks later.[22]

However, that was not the last of the legal proceedings arising from the case. A researcher in the 1990s made the surprising discovery that the hair samples collected at the time of the girl's death were still in the police archives, and re-testing was done by both the Victorian Institute of Forensic Medicine and the Australian Federal Police laboratory. This confirmed that the hair found on blankets in the defendant's home did not match the scalp sample of the dead girl. The Victorian Attorney-General referred the matter to the Supreme Court in 2007, which concluded unanimously that the conviction could not stand.[23] Crucial to this finding was a report by Dr James Robertson, then Director of Forensic Services at the Australian Federal Police, and an expert in forensic hair comparisons, whose analysis concluded that 'the hairs recovered from the brown-grey blanket could not have come from the deceased, Tirtschke'.[24] Relatives of both the prisoner and the victim signed a petition for mercy, and the Governor

22 The High Court decision is dated 5 April 1922, and Ross was hanged on 24 April 1922.

23 *Re Colin Campbell Ross* [2007] VSC 572 (20 December 2007) (Teague, Cummins and Coldrey JJ).

24 The forensic report of Dr James Robertson is included in full in the Supreme Court's judgment (at [80]), in recognition of its importance in resolving the case.

of Victoria posthumously pardoned Colin Campbell Ross in May 2008, some 86 years after his hanging.[25]

The Chamberlains

The *Chamberlain* case has been highly influential on the role of forensics in criminal proceedings. After two coronial inquiries into the 1980 disappearance of their baby daughter, Azaria, from a camping ground near Uluru in central Australia, Lindy and Michael Chamberlain were committed to stand trial in the Supreme Court of the Northern Territory. The prosecution case was that Lindy had killed Azaria in the family car and the couple had disposed of the body. The defence argued at trial that Azaria had been taken by a dingo. The prosecution case relied heavily on forensics, and after a highly publicised jury trial, Lindy was convicted of murder with Michael convicted as an accessory.

Appeals to higher courts were unsuccessful.[26] Continuing public disquiet with the convictions led to the establishment of a Royal Commission in 1987, which found profound flaws in the forensic evidence adduced by the prosecution. This included an alleged bloody handprint on Azaria's clothing, an expert's purported identification of damage to the clothing as caused by scissors rather than dingo teeth and, most critically, the identification of supposed foetal blood under the dashboard of the car. This was systematically discredited by the Commissioner, who observed that:[27] 'evidence was given at trial by experts who did not have the experience, facilities or resources necessary to enable them to express reliable opinions on some of the novel and complex scientific issues which arose for consideration'.

The Northern Territory Supreme Court, sitting as a Court of Criminal Appeal and acting on recommendations of the Morling Commission, quashed the convictions in 1988.[28] However, the cause of death was not officially determined to be due to a dingo taking Azaria until a fourth coronial inquest was completed in 2012.[29] The legacy of the Chamberlain saga is arguably that courts have become more willing to scrutinise forensic evidence, that forensic experts have improved

25 J. Silvester. (2008). Ross cleared of murder nearly 90 years ago. *The Age*. Retrieved from http://www.theage.com.au/news/national/bcrimeb-man-cleared-of-murder-86-years-after-he-was-executed/2008/05/26/1211653938453.html

26 *Re Alice Lynne Chamberlain and Michael Leigh Chamberlain v R* (1983) 72 FLR 1; *Chamberlain v R (No. 2)* (1984) 153 CLR 521. The High Court appeal failed with a 3:2 majority upholding the conviction.

27 *Report of the Commissioner the Hon. Mr. Justice T.R. Morling / Royal Commission of Inquiry into Chamberlain Convictions* (1987), 340–1, cited by Warren (2009).

28 *Reference under s.433A of the Criminal Code by the Attorney-General for the Northern Territory of Australia of Convictions of Alice Lynne Chamberlain and Michael Leigh Chamberlain* [1988] NTSC 64 (15 September 1988). Both Lindy and Michael Chamberlain were pardoned in 1987, though this did not legally overturn the convictions.

29 *Inquest into the death of Azaria Chantel Loren Chamberlain* [2012] NTMC 020. The third inquest, held in 1995, had returned an open finding.

their processes and clarified that they must act impartially in assisting the court rather than the prosecution and that both the judicial system and extra-judicial means of review are arguably more willing to re-examine past cases to identify and correct miscarriages of justice.

Edward Splatt

Edward Splatt was convicted of murder in 1978 and spent six and a half years in prison before being released on the recommendation of a Royal Commission, which was followed by *an ex gratia* payment of $300,000. The case against him was circumstantial and largely based on scientific analysis of paint, wood, birdseed and biscuit particles collected at the crime scene. Upon reviewing the case, the Royal Commissioner concluded that it would be 'unjust and dangerous for the verdict to stand' (ALRC, 1985; Dioso-Villa, 2014). The main reasons for this conclusion were that the investigation and forensic analysis were conducted by the same police officers, so that there was a lack of scientific objectivity and a reluctance to consider exculpating rather than incriminating interpretations of the evidence.[30] Following this case, and the *Chamberlain* case in which South Australian forensic technicians were also involved, forensic procedures were significantly reviewed and reformed. In particular, expert guidelines now emphasise that:[31]

> The role of the expert witness is to provide relevant and impartial evidence in his or her area of expertise. An expert should never mislead the Court or become an advocate for the cause of the party that has retained the expert.

This requirement for impartiality is supported by modern best practice in forensic laboratories, including blind testing samples identified only by numbers and where the analyst has no detail on the police investigation or prosecution involved.

Alexander McLeod-Lindsay

Alexander McLeod-Lindsay came home from his work one day in 1964 to find his wife and son severely beaten. Both survived, and the wife described the attacker. However, police suspected McLeod-Lindsay, and developed a theory that he had slipped away from the hotel and returned there unnoticed after attacking his family. Blood on his jacket was said to be 'impact splatter' that was deposited during the attack. McLeod-Lindsay was convicted of attempted murder and served almost ten years in prison before being released. Despite appealing to higher courts for review, the convictions stood, despite expert scientists arguing that the blood on the jacket displayed clotting, and therefore was most likely deposited when

30 B. Littley. (2012). Someone got away with murder. *Adelaide Advertiser*, 27 January.
31 See, for example, Federal Court of Australia, *Expert Evidence Practice Note* (GPN-EXPT), 25 October 2016.

McLeod-Lindsay held his wife in his arms upon coming home to the horrific scene.[32] It was not until a further inquiry in 1990 that a final exoneration and compensation were awarded by the state.[33]

Frank Button

The leading example in Australia of a DNA-based exoneration is the case of Frank Button, in which the Queensland Court of Appeal quashed the defendant's rape conviction when presented with post-trial DNA analysis indicating that someone other than Button was the rapist. The lead judgment stated:[34]

> As I said in the course of argument, today is a black day in the history of the administration of criminal justice in Queensland. The appellant was convicted of rape by a jury and has spent some approximate 10 months in custody in consequence of that conviction. DNA testing carried out at the insistence of his lawyers after that jury verdict has now established that he was not the perpetrator of the crime in question, and indeed the recent DNA testing would appear to have identified some other person as the perpetrator of that crime. What is of major concern to this Court is the fact that that evidence was not available at the trial.
>
> What is disturbing is that the investigating authorities had also taken possession of bedding from the bed on which the offence occurred, and delivered those exhibits to the John Tonge Centre. No testing of that bedding was carried out prior to trial. The explanation given was that it would not be of material assistance in identifying the appellant as the perpetrator of the crime.

The Director of Public Prosecutions referred to a lack of adequate resourcing of the State's main forensic facility. However, the Court of Appeal observed:

> It may well be that laboratory testing is expensive, particularly if it is to be as extensive as in my view it should be, but the cost to the community of that testing is far less than the cost to the community of having miscarriages of justice such as occurred here. The cost to the community in a case like this includes not only the costs of both sides of the aborted trial, but the costs to the appellant of the fact that he has been in custody for the length of time …

32 *Report of the Inquiry held under Section 475 of the Crimes Act 1900 into the Conviction of Alexander McLeod-Lindsay*, 1969.

33 M. Brown, 'Exonerated 26 years after his conviction' (*Sydney Morning Herald*, 21 September 2009), written on the death of Alexander McLeod-Lindsay two days earlier.

34 *R v Button* [2001] QCA 133 (10 April 2001), (Williams JA, White and Holmes JJ concurring), perhaps Australia's only DNA-based exoneration appeal (Roach 2015). The judge's words were adopted by an Australian Broadcasting Corporation documentary about the case, 'A Black Day for Justice' (see discussion in Chapter 3 and in Smith 2015).

This case illustrates that forensic science, including biometrics, can only be of consistent and reliable assistance in criminal proceedings if analysis is conducted in a comprehensive and scientifically robust manner. The risk otherwise is that miscarriages of justice will occur, and they may not always be amenable to remedial justice through criminal appeals of other forms of post-conviction review.[35]

Ensuring the reliability of biometrics

A recurring theme in the cases discussed in this and the preceding chapter is the need for the use of biometric identification to be premised on reliable scientific techniques, applied in a consistent and verifiable manner in investigations. If substandard techniques or even 'junk science' are allowed into the process, then the results that follow may well be miscarriages of justice (Dioso-Villa et al., 2016). Historically, the role of dubious forensic analysis has been highlighted, as well as the fact that some processes and practices have been improved. This last topic explores reforms on a systematic basis that relate specifically to biometric identification.

Drawing on a landmark report into forensic science in the United States (NAS, 2009), commentators have identified the following as key problems affecting the use of biometrics (Ross, 2012; Roux, Crispino & Ribaux, 2012; Edmond, Martire & San Roque, 2011; Edmond, 2014, 2015):

- *Validation of scientific techniques*: While some new areas such as DNA identification have been reasonably well validated through court cases assessing their scientific basis, this is less true for newer techniques such as facial or body mapping;
- *Standard protocols*: Not all types of biometric identification operate according to clear and agreed processes for the collection and analysis of material e.g. voice identification may be based on *ad hoc* expertise rather than a standard approach across cases;
- *Inaccuracy and bias*: Conclusions that appear to be based on scientific analysis may not disclose matters affecting their accuracy, the language used may be highly technical without adding to the accuracy of the analysis, and sample biases may not be disclosed where they are known.

Reforms have tended to focus on the accreditation of scientific laboratories and training, with peer-reviewed research and validation required to be systematically employed for quality assurance (Ross, 2012). Some legal academics have argued for a greater judicial focus on reliability as a threshold requirement for the admissibility

35 Not discussed here are other noteworthy miscarriage of justice cases involving forensics, such as those involving John Button and Andrew Mallard in Western Australia, and Gordon Wood in New South Wales, as these cases did not rely on biometrics as the principal means of identification relied on by the prosecution.

of scientific evidence generally (Edmond, 2014; Ligertwood, 2015). In the context of the rules governing the admission or exclusion of evidence, this means ensuring that the relevance or probative value requirement, the rules allowing expert opinion and the use of discretionary exclusion based on unfair prejudice need to be applied carefully. The following assessment indicates that this is possible within existing rules when appropriately interpreted (Ligertwood, 2015):

> First, the admissibility rules relating to relevance, opinion and discretion are open to interpretations permitting the rigorous consideration of forensic evidence, to ensure that it is based on theoretical and/or empirical grounds and that it is expressed transparently in a way that enables the trier of fact, with appropriate directions from the trial judge, to take it rationally into account when considering the criminal standard of proof.
>
> Secondly, standards governing appellate review (including post-conviction review) are open to interpretations that could ensure that forensic evidence is carefully scrutinised on appeal, not only to determine its admissibility and use but also in determining whether the criminal standard of proof has been satisfied. Thirdly, the adversary process may be limited by time and resources but it undoubtedly has the potential to provide a powerful scrutiny of forensic evidence.
>
> And finally, as far as the common lack of scientific expertise among the judges and lawyers who must try to comprehend and evaluate forensic evidence is concerned, one might argue that in many cases it is not necessary for laypersons (judges and juries) to follow all the technicalities of a forensic process and it is enough to appreciate the possibilities of error in determining admissibility and proof. It is only where the very basis of scientific evidence is being disputed that persons with a background in that area of science may be required to adjudicate the dispute.

This suggests that it is within the capacity of the legal system, assisted by the forensic sciences, to make the best use of biometrics in the courtroom, in criminal trials and appeals, and in other forms of post-conviction review.

References

Auld, R. (2001). *Review of the Criminal Courts of England and Wales*. London: UK Stationery Office.

Australian Capital Territory Department of Justice and Community Safety (ACT JACS). (2015). *Double Jeopardy Information Paper*. Retrieved from http://www.justice.act.gov.au/review/view/38/title/double-jeopardy-information-paper

Australian Law Reform Commission (ALRC). (2003). *Essentially Yours: The Protection of Human Genetic Information in Australia Report 96*. Retrieved from http://www.austlii.edu.au/au/other/lawreform/ALRC/2003/96.html

Australian Law Reform Commission (ALRC). (1985). Compensation for imprisonment. *Australian Law Reform Commission Reform Journal 39*, 105.

Burton, K. (2004). Reform of the double jeopardy rule on the basis of fresh and compelling evidence in New South Wales and Queensland. *James Cook University Law Review 11*, 84.

Caruso, D. & Crawford, N. (2014). The executive institution of mercy in Australia: The case and model for reform. *University of New South Wales Law Journal 37*(1), 312.

Christian, K. (2001). And the DNA shall set you free: Issues surrounding post-conviction DNA evidence and the pursuit of innocence. *Ohio State Law Journal 62*, 1195.

Corns, C. (2003). Retrial of acquitted persons: Time for reform of the double jeopardy rule? *Criminal Law Journal 27*, 80.

Corns, C. & Urbas, G. (2008). Criminal appeals 1907–2007: Issues and perspectives. *Law in Context 26*(1), 1.

Cowdery, N. (2005). Wrongful conviction and double jeopardy. *Judicial Officers' Bulletin. 17*(4), 27.

De Foore, D. (2002). Postconviction DNA testing: A cry for justice from the wrongly convicted. *Texas Technical Law Review 13*(2), 491.

Dioso-Villa, R. (2014). Out of grace: Inequity in post-exoneration remedies for wrongful conviction. *University of New South Wales Law Journal 37*(1), 349.

Dioso-Villa, R., Julian, R., Kebbell, M., Weathered, L. & Westera, N. (2016). Investigation to exoneration: A systemic review of wrongful conviction in Australia. *Current Issues in Criminal Justice 28*(2), 157.

Edmond, G. (2014). The 'science' of miscarriages of justice. *University of New South Wales Law Journal 37*(1), 376.

Edmond, G. (2015). What lawyers should know about the forensic 'sciences'. *Adelaide Law Review 36*(1), 33.

Edmond, G., Martire, K., & San Roque, M. (2011). Unsound law: Issues with ('expert') voice comparison evidence. *Melbourne University Law Review 35*(1), 52.

Edmond, G. & Roberts, A. (2011). Principles of evidence law and their implications for forensic science and medicine. *Sydney Law Review 33*, 359.

Goldingham, C. (2002). Miscarriage of justice: And the truth will set you free. *Alternative Law Journal 27*(3), 140.

Griffith, G. & Roth, L. (2006). *DNA Evidence, Wrongful Convictions and Wrongful Acquittals-* Sydney: Parliament of New South Wales.

Hamer, D. (2014). Wrongful convictions, appeals, and the finality principle: The need for a criminal cases review commission. *University of New South Wales Law Journal 37*(1), 270.

Hamer, D. (2015). The Eastman case: Implications for an Australian criminal cases review commission. *Flinders Law Journal 17*, 433.

Hamer, D. & Edmond, G. (2013). Truth or lies: Overturning wrongful convictions. *The Conversation*. Retrieved from http://theconversation.com/truth-or-lies-overturning-wrongful-convictions-20430

Ligertwood, A. (2011). Can DNA evidence alone convict an accused? *Sydney Law Review 33*(3), 487.

Ligertwood, A. (2015). What lawyers should and can do now that they know about the forensic sciences. *Adelaide Law Review 36*(1), 153.

Milne, S. (2015). The second or subsequent criminal appeal, the prerogative of mercy and the judicial inquiry: The continuing advance of post-conviction review. *Adelaide Law Review 36*(1), 211.

Model Criminal Code Officers Committee of the Standing Committee of Attorneys-General (MCCOC). (2003). *Model Criminal Code. Chapter 2, Issue Estoppel, Double Jeopardy and Prosecution Appeals against Acquittals: Discussion Paper.* Canberra: Australian Government.

National Research Council of the National Academy of Sciences (NAS). (2009). *Strengthening Forensic Science in the United States: A Path Forward.* Washington, DC: National Academies Press.

Penhallurick, C. (2003). The proviso in criminal appeals. *Melbourne University Law Review. 27*(3), 800.

Roach, K. (2015). Comparative reflections on miscarriages of justice in Australia and Canada. *Flinders Law Journal 17*(2), 381.

Roux, C., Crispino, F. & Ribaux, O. (2012). From forensics to forensic science. *Current Issues in Criminal Justice 24*(1), 7.

Ross, A. (2012). Forensic science in Australia: Where does Australia sit in relation to trends and issues in the international context? *Current Issues in Criminal Justice 24*(1), 121.

Rudstein, D. (2014). *Double Jeopardy: A Reference Guide to the United States Constitution*Westport, CT: Praeger Publishers.

Sangha, B. (2015). The statutory right to second or subsequent criminal appeals in south Australia and Tasmania. *Flinders Law Journal 17*, 471.

Sangha, B. & Moles, R. (2012). Post-appeal review rights, Australia, Britain and Canada. *Criminal Law Journal 36*, 300.

Sangha, B. & Moles, R. (2015). *Miscarriages of Justice: Criminal Appeals and the Rule of Law in Australia.* Sydney: LexisNexis.

Smith, M. (2016). *DNA Evidence in the Australian Legal System.* Chatswood, NSW: LexisNexis Butterworths.

Spencer, J. (1982). Criminal appeals: The tail that wags the dog. *Criminal Law Review 3*, 260.

Thomas, G. (1998). *Double Jeopardy: The History, The Law.* New York: New York University Press.

Urbas, G. (2002). DNA evidence in criminal appeals and post-conviction inquiries: Are new forms of review required? *Macquarie Law Journal 2*, 141.

Urbas, G. (2012). Case note on Bui v DPP (Cth) – The High Court considers double jeopardy in sentencing appeals. *University of Notre Dame Australia Law Review 14*, 187.

Warren, M. (2009). Remarks on the occasion of the 19th International Symposium on the Forensic Sciences. *Monash University Law Review 35*(1), 19–24.

Weathered, L. (2003). Investigating innocence: The emerging role of innocence projects in the correction of wrongful conviction in Australia. *Griffith Law Review 12*(1), 64.

Weathered, L. (2004). A question of innocence: Facilitating DNA-based exonerations in Australia. *Deakin Law Review 9*(1), 279.

Weathered, L. (2013). Reviewing the New South Wales DNA review panel: Considerations for Australia. *Current Issues in Criminal Justice. 24*(3): 449.

Weathered, L. & Blewer, R. (2009). Righting wrongful convictions with DNA innocence testing: Proposals for legislative reform in Australia. *Flinders Journal of Law Reform 11*(1), 43.

Young, W. (2010). The role of the courts in correcting miscarriages of justice. *Canterbury Law Review 16*, 256.

Zdenkowski, G. (1993). Remedies for miscarriage of justice. *Current Issues in Criminal Justice 5*(1), 105.

INDEX

Note: page numbers in **bold** refer to text within tables.